Sciences
Engineering
Medicine

CW00537687

Operationalizing Sustainable Development to Benefit People and the Planet

Committee on Operationalizing
Sustainable Development

Science and Technology for
Sustainability Program

Policy and Global Affairs

Consensus Study Report

NATIONAL ACADEMIES PRESS 500 Fifth Street, NW Washington, DC 20001

This activity was supported by Academia Sinica, Arizona State University's Julie Ann Wrigley Global Futures Laboratory, Carnegie Corporation of New York under award number G-21-58294, Elsevier, JPB Foundation under award number 2020-1963, the Peace Department, and the George and Cynthia Mitchell Endowment for Sustainability. Any opinions, findings, conclusions, or recommendations expressed in this publication do not necessarily reflect the views of any organization or agency that provided support for the project.

International Standard Book Number-13: 978-0-309-69165-9
International Standard Book Number-10: 0-309-69165-6
Digital Object Identifier: https://doi.org/10.17226/26654

This publication is available from the National Academies Press, 500 Fifth Street, NW, Keck 360, Washington, DC 20001; (800) 624-6242 or (202) 334-3313; http://www.nap.edu.

Printed in the United States of America.

Suggested citation: National Academies of Sciences, Engineering, and Medicine. 2022. *Operationalizing Sustainable Development to Benefit People and the Planet.* Washington, DC: The National Academies Press. https://doi.org/10.17226/26654.

The **National Academy of Sciences** was established in 1863 by an Act of Congress, signed by President Lincoln, as a private, nongovernmental institution to advise the nation on issues related to science and technology. Members are elected by their peers for outstanding contributions to research. Dr. Marcia McNutt is president.

The **National Academy of Engineering** was established in 1964 under the charter of the National Academy of Sciences to bring the practices of engineering to advising the nation. Members are elected by their peers for extraordinary contributions to engineering. Dr. John L. Anderson is president.

The **National Academy of Medicine** (formerly the Institute of Medicine) was established in 1970 under the charter of the National Academy of Sciences to advise the nation on medical and health issues. Members are elected by their peers for distinguished contributions to medicine and health. Dr. Victor J. Dzau is president.

The three Academies work together as the **National Academies of Sciences, Engineering, and Medicine** to provide independent, objective analysis and advice to the nation and conduct other activities to solve complex problems and inform public policy decisions. The National Academies also encourage education and research, recognize outstanding contributions to knowledge, and increase public understanding in matters of science, engineering, and medicine.

Learn more about the National Academies of Sciences, Engineering, and Medicine at **www.nationalacademies.org**.

COMMITTEE ON OPERATIONALIZING
SUSTAINABLE DEVELOPMENT

E. WILLIAM COLGLAZIER (*Co-Chair*), Editor-in-Chief, *Science & Diplomacy*, and Senior Scholar, Center for Science Diplomacy, American Association for the Advancement of Science

CHERRY MURRAY (NAS/NAE) (*Co-Chair*), Professor of Physics and Deputy Director for Research, Biosphere 2, The University of Arizona

ERIN BROMAGHIM, Deputy Mayor of International Affairs, Mayor's Office of International Affairs, City of Los Angeles

HARINI NAGENDRA, Director, Professor, and Lead, Centre for Climate Change and Sustainability, Azim Premji University

NEBOJSA NAKICENOVIC, Executive Director, The World In 2050

ILONA M. OTTO, Professor, Societal Impacts of Climate Change, Wegener Center for Climate and Global Change, University of Graz

ALFRED WATKINS, Founder and Chairman, Global Solutions Summit

Science and Technology for Sustainability Program Staff

FRANKLIN CARRERO-MARTÍNEZ, Senior Director, Global Policy and Development and Science and Technology for Sustainability

EMI KAMEYAMA, Program Officer

DANIELLE ETHERIDGE, Administrative Assistant

Consultant

PAULA TARNAPOL WHITACRE, Principal, Full Circle Communications, LLC

Preface and Acknowledgments

In February 2022, the National Academies of Sciences, Engineering, and Medicine's Science and Technology for Sustainability Program initiated a new study, *Operationalizing Sustainable Development*. This followed a series of discussions held by the National Academies Roundtable on Science and Technology for Sustainability that explored the importance of U.S. engagement on the United Nations Sustainable Development Goals (SDGs) and a high-level convening during the Nobel Prize Summit *Our Planet, Our Future* on April 26-28, 2021, that focused on pressing global challenges: climate change and biodiversity loss, rising inequality, and rapid societal transformation enabled by emerging and converging technologies. While the pandemic and heightened geopolitical conflicts have made the achievement of the SDGs more challenging, there is a lack of shared understanding of how to operationalize sustainable development and accelerate the pace of global policy discussions. To address this challenge, a committee with a range of expertise and experience in government, academia, business, and nongovernmental organizations was convened. Brief biographies of the individual committee members are provided in Appendix A. At this moment, with many heightened concerns regarding sustainability issues, the committee was charged to produce a *short* consensus report that identifies key research priorities and possible actionable steps for operationalizing sustainable development. Although the report's findings are directed toward U.S. stakeholders to consider, the committee believes they can be used by policy makers, researchers, managers, education administrators, and practitioners in the United States and globally to make a measurable difference in a sustainable future for all.

During the course of the study, the committee conducted two information-gathering workshops and committee meetings. The two public workshops explored positive case studies for operationalizing sustainable development in eight theme areas: (1) Education and Capacity Building; (2) Localization of the SDGs and Indigenous Knowledge Networks; (3) Sustainable and Equitable Food Systems; (4) Urbanization; (5) Decarbonization; (6) Science, Technology, and Innovation; (7) Science and Peace; and (8) Financing for Sustainable Development. An agenda for each workshop is provided in Appendix B. The session on science, technology, and innovation was held as part of the side event of the 7th Multi-stakeholder Forum on Science, Technology and Innovation for the Sustainable Development Goals (STI Forum).

The committee gratefully acknowledges the following individuals for making presentations at its information-gathering workshops: Jean-Paul Adam, United Nations Economic Commission for Africa; Stephanie Arcusa, Arizona State University; Aparna Basnyat, United Nations Development Programme; Prateek Bumb, Carbon Clean; Erin Burns, Carbon180; Celeste Connors, Hawai'i Green Growth; Kathleen Draper, International Biochair Initiative; Niki Frantzeskaki, Utrecht University; Alison Grantham, Grow Well Consulting, LLC; Adam Roy Gordon, United Nations Global Compact; Ravi Kanbur, Food System Economics Commission and Cornell University; Ashish Kothari, Kalpavriksh; Marianna Koval, New York University; Lykke Leonardsen, City of Copenhagen; Jannie Lilja, Stockholm International Peace Research Institute; Rachel Locke, University of San Diego; Mark McCaffrey, The Long Game; Sarah Mendelson, Carnegie Mellon University; Eduard Müller Castro, University for International Cooperation; Jan Marco Müller, European External Action Service; Lauren Muusse, World Benchmarking Alliance; Carol O'Donnell, Smithsonian Institution; Sabine O'Hara, University of the District of Columbia; Susan Parnell, University of Cape Town; Anthony Pipa, Brookings; Emilia Saiz, United Cities and Local Governments; Worajit Setthapun, Chiang Mai Rajabhat University; Roman Słowiński, Polish Academy of Sciences; Atsushi Sunami, Sasakawa Peace Foundation; Meghna Tare, The University of Texas, Arlington; Klaus Tilmes, Senior Policy Advisor and Development Consultant; and Marc Weiss, Global Urban Development. The information provided at these workshops is used throughout this report as the primary source of information from which the key research priorities and possible actionable steps were developed.

The report would not have been possible without the sponsors of this study, including Academia Sinica, Arizona State University's Julie Ann Wrigley Global Futures Laboratory, Carnegie Corporation of New York, Elsevier, JPB Foundation, The Peace Department, and the George and Cynthia Mitchell Endowment for Sustainability. The committee gratefully acknowledges Dana Bourland, JPB Foundation; Amanda Ellis and Jessica Givens, Arizona State University; Ann Gabriel, Elsevier; and James Sternlicht, The Peace Department for making presentations to the committee.

This Consensus Study Report was reviewed in draft form by individuals chosen for their diverse perspectives and technical expertise. The purpose of this independent review is to provide candid and critical comments that will assist the National Academies of Sciences, Engineering, and Medicine in making each published report as sound as possible and to ensure that it meets the institutional standards for quality, objectivity, evidence, and responsiveness to the study charge. The review comments and draft manuscript remain confidential to protect the integrity of the deliberative process.

We wish to thank the following individuals for their review of this report: Arun Agrawal, University of Michigan; Mark Barteau, Texas A&M University; Anthony Chase, Occidental College; David Dzombak, Carnegie Mellon University; Garrick Louis, University of Virginia; Sarah Mendelson, Carnegie Mellon University; and Susan Parnell, University of Bristol.

Although the reviewers listed above have provided many constructive comments and suggestions, they were not asked to endorse the conclusions or recommendations, nor did they see the final draft of the report before its release. The review of this report was overseen by David Allen, The University of Texas at Austin. He was responsible for making certain that an independent examination of this report was carried out in accordance with the standards of the National Academies and that all review comments were carefully considered. Responsibility for the final content rests entirely with the authoring committee and the National Academies.

The committee is also grateful for the assistance of the National Academies staff in organizing this report. Staff members who contributed to this effort are Franklin Carrero-Martínez, senior director, Science and Technology for Sustainability Program; Emi Kameyama, program officer; and Paula Whitacre, consultant.

Finally, we especially thank the members of the committee for their tireless efforts throughout the development of this report.

E. William Colglazier and Cherry Murray, Co-Chairs
Committee on Operationalizing Sustainable Development

Contents

Executive Summary

The COVID-19 pandemic and overlapping global crises, including geopolitical conflict and climate change, have made achievement of the United Nations Sustainable Development Goals (SDGs) more challenging. The scientific community increasingly recognizes the need to accelerate the adoption of evidence-based, scientifically-sound policies and actions to operationalize sustainable development. Achieving the SDGs will require broad engagement and commitment from governments, the private sector, funders, and civil society (ISC, 2020); however, stakeholders lack a shared understanding of how the 17 SDGs can be operationalized (Sachs et al., 2019). Moreover, despite the high degree of interest in the types of activities included in the SDGs, recognition of the SDGs is low in the United States (Morning Consult and United Nations Foundation, 2021).

The National Academies Committee on Operationalizing Sustainable Development was charged with identifying key research priorities and possible actionable steps to operationalize sustainable development at the global and local levels. The committee convened two virtual public workshops to gather information on positive case studies across eight interrelated themes (summarized below), which served as the primary source of evidence for its work. The committee also drew on global case studies and papers, previous National Academies activities such as the Nobel Prize Summit 2021, a bounded literature review, and members' expertise. Although the scope of the challenges and opportunities are global with many research investigations and actions needed, the committee developed the following research priorities and possible actionable steps for consideration by U.S. stakeholders, informed mostly by the workshops:

1. **Education and Capacity Building (Chapter 2):** Education is critical to achieving the SDGs, and educational institutions at all levels are powerfully positioned to operationalize sustainable development across society. Universities could undertake initiatives to assist faculty and students to develop Voluntary University Reviews to ensure that students regardless of major are exposed to the challenges and opportunities in sustainable development, and to partner with local and national government and nonprofit organizations to advance the SDGs. Cities and school districts could initiate and support locally relevant K-12 learning on the SDGs, and governments and education leaders could engage the public to raise awareness of the SDGs.

2. **Localization of the SDGs and Indigenous Knowledge Networks (Chapter 3):** The SDGs embrace global aspirations, but they must be rooted in local buy-in and implementation. Urban and community leaders and practitioners as well as philanthropic organizations could learn from case studies and knowledge networks, including how others effectively incorporate indigenous knowledge to advance sustainability. The U.S. government could commit to creating a Voluntary National Review (VNR) by encouraging more states and cities to conduct Voluntary Local Reviews and synthesize this already good work at the local level to scale to a VNR roll-up.

3. **Food Systems (Chapter 4):** The food system encompasses a wide range of activities from input supply and the production of crops, livestock, fish, and other agricultural commodities to storage, transportation, processing, packaging, consumption, and waste disposal. The current food system is responsible for one-third of global greenhouse gas emissions and 70 percent of global water use. Although it produces an abundance of food (as well as food waste and ecological damage), approximately 2.4 billion people (or 30 percent of the global population) lacked access to adequate food in 2020, even before the current rise in global food prices and the disruption of the supply chain exacerbated the situation. Addressing these issues and operationalizing the SDGs will entail holistic reform of each link in the food system, taking into account environmental, scientific, economic, and social factors.

4. **Urbanization (Chapter 5):** Local-scale sustainability transformations are important, and they are easier to achieve than systemic change across multiple dimensions of the SDGs at the city-regional scale. It is important to assess what kinds of big data are needed from cities to monitor SDG transitions along environmental, social, and economic considerations and to demonstrate the multiple benefits of sustainable urbanization.

5. **Decarbonization (Chapter 6):** Decarbonization of energy systems is central to global decarbonization efforts and achievement of all SDGs. A fundamental energy-systems transformation would help to address health, climate, and other challenges facing humanity, and would especially benefit individuals without access to affordable and clean energy services. In addition to reducing net emissions across all sectors and transitioning to zero-carbon energy sources, both organic and inorganic carbon dioxide removal (CDR) will minimize global warming. Needed are massive scale-up, financial investment, and research that examine the fundamental science of ocean- and nature-based CDR; the shift toward zero-carbon sources of energy such as renewables and nuclear; efficiency improvement across the whole energy system especially in energy end use; and new behaviors and lifestyles including circularity and sufficiency.

6. **Science, Technology, and Innovation for the SDGs (Chapter 7):** Partnerships across sectors and disciplines, including science, technology, and innovation (STI), offer hope for resurgent multilateralism and innovative approaches to advance the SDGs. Digital and other technological advances offer new possibilities across sectors and communities. Studies could examine the current status of achieving the SDGs and how STI can facilitate their achievement in the context of the economic crisis, the COVID-19 pandemic, and conflicts including the war in Ukraine.

7. **Science and Peace (Chapter 8):** SDG 16 covers a lot of ground including the reduction of all forms of violence, equal access to justice for all, increased accountability and transparency, and the protection of fundamental freedoms. Conflict undermines achievement of all SDGs. There is a need to strengthen SDG data hubs, partnerships, and data for monitoring and enforcement related to the reduction of violence aspects with a focus on science and peace. Governments and nongovernmental organizations could create a new global social pact, as part of the effort to negotiate the initiative that follows the SDGs beyond 2030, to promote a global system for science to advance peace.

8. **Financing to Achieve the SDGs (Chapter 9):** Despite the challenges, opportunities exist to realize tangible and intangible benefits from SDG-related investing. Place-based initiatives could help to attract private investment and unlock financing. Public, private, and other organizations could create more blended finance options given the growing demand for positive environmental, social, and governance investments.

Chapters 2–9 list all of the research priorities and possible actionable steps. Chapter 10 then compiles and organizes all of the committee's research priorities and possible actionable steps that can help to operationalize sustainable development by stakeholder. The committee believes that these recommendations are ambitious but realistic and, taken together, can make a measurable difference in a sustainable future for all.

1

Introduction

The stresses on the planet caused by human activity over the past century, and especially the past few decades, are felt in different ways but are evident worldwide. Accelerated threats to wildlife and plant species, soil, air, and water not only harm the natural world, but also affect current and future human livelihoods and well-being. Biodiversity loss, climate change, species' migration patterns, and more are adversely impacting people in the United States and around the world—and disproportionately impacting people living in poverty or who are otherwise marginalized (IPCC, 2022; UNDP, 2020). The 1987 United Nations (UN) report *Our Common Future*, known as the Brundtland report, defines sustainable development as "development that meets the needs of the present without compromising the ability of future generations to meet their own needs" (WCED, 1987). The report emphasizes three components of sustainable development: environmental protection, economic growth, and social equity.

When 193 national leaders unanimously adopted the Sustainable Development Goals (SDGs) in 2015 (Figure 1-1), they did so with the expressed hope of providing a "shared blueprint for peace and prosperity for people and the planet, now and into the future." The 17 goals and their 169 targets embrace economic growth, social inclusion, and environmental protection. Interlinkages across these goals and targets generate not only multiple synergies but also tradeoffs (i.e., progress on one goal supports or could hinder progress toward other goals). With the SDGs already a challenge to achieve by the designated time target of 2030, the COVID-19 pandemic and heightened conflict worldwide have disrupted fragile progress and sharpened existing inequities.

FIGURE 1-1 United Nations Sustainable Development Goals.
SOURCE: United Nations, 2019. Communications materials. https://www.un.org/sustainable
development/news/communications-material.

Despite setbacks and stagnation, from small villages to mega-cities, people are joining together to operationalize sustainable development, that is, to identify and create solutions to advance the well-being of people and the planet, both within and between communities around the world. In addition to often overcoming technological and scientific obstacles, the successes reveal the imperative to change behaviors embedded in culture, norms, and values, as well as to link policies with on-the-ground actions. Many observers see a time horizon of a few years, or at most decades, to make meaningful change to avert what could become an ecological catastrophe. Governing bodies and the multilateral system wrestle with constraints that slow their response to urgent transnational challenges. Yet, a wide swath of society—including scientists, youth, indigenous communities, marginalized groups, artists, faith leaders, and others—are not waiting. They are taking action.

In April 2021, the Nobel Prize Summit *Our Planet, Our Future* captured these diverse voices and perspectives. Hosted by the Nobel Foundation and organized by the National Academies of Sciences, Engineering, and Medicine, Potsdam Institute for Climate Change Research, and Stockholm Resilience Centre, the 3-day summit featured Nobel Laureates interacting with scientists, artists, community activists, policy makers, and other stakeholders (NASEM, 2021a). Although not charged with issuing formal recommendations, the summit attendees created a powerful output: a Call to Action signed by 126 Nobel Laureates to present to leaders of the G-7 countries and the UN Secretary General, among others. The Call to Action states:

> Global sustainability offers the only viable path to human safety, equity, health, and progress. Humanity is waking up late to the challenges and opportunities of active planetary stewardship. But we *are* waking up. Long-term, scientifically based decision-making is always at a disadvantage in the contest with the needs of the present. Politicians and scientists must work together to bridge the divide between expert evidence, short-term politics, and the survival of all life on this

planet in the Anthropocene epoch. The long-term potential of humanity depends upon our ability today to value our common future. Ultimately, this means valuing the resilience of societies and the resilience of Earth's biosphere.[1]

FRAMEWORKS FOR ANALYSIS AND ACTION

The SDGs, also known as the Global Goals, "recognize that action in one area will affect outcomes in others and that development must balance social, economic, and environmental sustainability" (UNDP, 2022b). Since the adoption of the SDGs during the UN General Assembly in September 2015, many researchers, practitioners, policy makers, and the public have framed their efforts around the 17 goals and their 169 targets. The field of sustainability science has deepened over the past several decades, bringing together scientists, engineers, humanists, and others (NASEM, 2021b), and sustainability education is a growing multidisciplinary program of study in a growing number of colleges and universities (NASEM, 2020). During a July 2021 workshop, participants in a National Academies Roundtable on Science and Technology for Sustainability suggested that increasing awareness of the SDGs, with a focus on youth, civil society, and local governments, can serve to align activities for greater effect. Despite the high degree of interest in the types of activities explicitly or implicitly included in the SDGs, awareness of the SDGs is low in the United States (Morning Consult and United Nations Foundation, 2021).

Although the SDGs are interrelated and mutually reinforcing, complementary frameworks, such as The World in 2050 (TWI2050, 2018) and Doughnut Economics (Raworth, 2017), help us to better understand their interdependencies and tradeoffs. In addition to the contributions of experts across disciplines, the interconnection between goals is best understood through the lived experience of individuals. As the Committee on Operationalizing Sustainability was reminded during its first meeting that culminated in this report, "People living in poverty are suffering disproportionately from the harmful effects of unsustainable development and climate change. It's . . . exactly these people who have the solutions we need to be sitting down with and listening to, and implementing their visions" (Bourland, 2022). The committee also took note of indigenous knowledge and practices, especially those related to the natural world. The committee concurred with several comments made during its public workshops that, beyond sustainability, the goals should be to regenerate, not just sustain, and to thrive, not just to grow (Schueman, 2021). Similar to those who refute hopes for "getting back to normal" in the post-COVID-19 world with the observation that pre-pandemic normalcy had deep problems, many observers point out that "sustaining" the current situation is insufficient. A workshop participant shared a proverb from Hawai'i that captures this concept: A beach should not be left in the condition in which a person finds it—it should be left in a better condition (Connors, 2022).

---

[1] The full Call to Action can be read at https://www.nationalacademies.org/news/2021/04/nobel-prize-laureates-and-other-experts-issue-urgent-call-for-action-after-our-planet-our-future-summit.

THE CURRENT STUDY

The National Academies of Sciences, Engineering, and Medicine's Science and Technology for Sustainability (STS) Program serves as the institution's focus to harness the power of science, engineering, and medicine to meet sustainability challenges from local to global scales. Through STS, the National Academies convened an expert committee to identify key research priorities and possible actionable steps needed to operationalize sustainable development at the global and local levels (see Box 1-1 for the Statement of Task and Appendix A for biographical sketches of the committee members).

As noted in the Statement of Task, the committee was charged with convening two public workshops on which to draw findings for its consensus report. It should be underscored that the charge was for a *brief* report, with *key research priorities* and *possible actionable steps*. In reviewing its charge, the committee agreed that it would draw on global case studies and papers, but its recommendations would be directed toward U.S. stakeholders. Intended audiences are local, state, and federal government policy makers and managers; higher education administrators and researchers; research funders in the public, private, and nonprofit spaces; and civil society institutions and the public. Agreeing with the need for urgency and practical application, the committee adopted an ambitious timeline of less than 6 months to address its task. As input, the committee organized a series of virtual workshops to gather information on case studies across sectors and levels. With an almost limitless array of possible topics, the committee found that several reports, such as *The World in 2050* (TWI2050, 2018), *Six Transformations* (Sachs et al., 2019,

---

BOX 1-1
Committee Statement of Task

An ad hoc committee of the National Academies of Sciences, Engineering, and Medicine will identify key research priorities and possible actionable steps needed to operationalize sustainable development at the global and local levels. The committee will convene two virtual public workshops focused on (1) global and (2) local strategies as the primary source of evidence for its work, supplemented by background materials collected for the workshops and discussions held at the April 2021 Nobel Prize Summit and the July 2021 Sustainability Roundtable meeting. Each workshop will focus on pressing global challenges, such as science, technology, and innovation to enable societal transformations, including strengthening the science-policy interface, promoting innovation cohesion, and creating research agendas to inform post-2030 processes for the United Nations Sustainable Development Goals (SDGs). Following the workshops, the committee will prepare a brief consensus report that will identify key research priorities and possible actionable steps needed to operationalize sustainable development in the specific areas discussed at the workshops.

TABLE 1-1 List of Topics in Relevant Reports on Sustainable Development

The World in 2050 (TWI2050, 2018)	*Six Transformations* (Sachs et al., 2019)	*Global Sustainable Development Report 2019* (UN, 2019b)	*The European Green Deal* (EC, 2019)
Human capacity and health	Education, gender, and inequality	Human well-being and capabilities	A zero pollution ambition for a toxic-free environment
Consumption and production	Health, well-being, and demography	Sustainable and just economies	Preserving and restoring ecosystems and biodiversity
Decarbonization and energy	Energy decarbonization and sustainable industry	Food systems and nutrition patterns	From "Farm to Fork:" a fair, healthy, and environmentally friendly food system
Food, biosphere, and water	Sustainable food, land, water and oceans	Energy decarbonization and universal access	Accelerating the shift to sustainable and smart mobility
Smart cities	Sustainable cities and communities	Urban and peri-urban development	Building and renovating in an energy- and resource-efficient way
Digital revolution	Digital revolution for sustainable development	Global environmental commons	Mobilizing industry for a clean and circular economy
			Supplying clean, affordable, and secure energy
			Increasing the EU's climate ambition for 2030 and 2050

SOURCE: Committee generated, based on EC, 2019; Sachs et al., 2019; TWI2050, 2018; UN, 2019b.

drawing on TWI2050, 2018), *Global Sustainable Development Report 2019* (UN, 2019b), and *The European Green Deal* (EC, 2019), provided a useful framework for organizing the sessions (Table 1-1).

Based on a review of these reports and a series of discussions, the committee identified eight themes for the public workshops convened in April and May 2022:[2] (1) Education and Capacity Building; (2) Localization of the SDGs and Indigenous Knowledge Networks; (3) Food Systems; (4) Urbanization; (5) Decarbonization; (6) Science, Technology, and Innovation; (7) Science and Peace; and (8) Financing to Achieve the SDGs (Table 1-2). The committee

---

[2] See agendas in Appendix B. Workshop videos and selected presentations are available at https://www.nationalacademies.org/our-work/operationalizing-sustainable-development.

TABLE 1-2 Eight Themes in This Report and Their Key Relevant SDGs

Themes in This Report	Key Relevant SDGs
Education and Capacity Building	Goals 4 (quality education), 5 (gender equality), 8 (decent work), and 17 (partnerships for the goals)
Localization of the SDGs and Indigenous Knowledge Networks	Goals 1 (no poverty), 10 (reduced inequalities), 11 (sustainable cities and communities), and 17 (partnerships for the goals)
Food Systems	Goals 1 (no poverty), 2 (zero hunger), 3 (good health and well-being), 6 (clean water), 10 (reduced inequalities), 12 (responsible consumption and production), 13 (climate action), 14 (life below water), and 15 (life on land)
Urbanization	Goals 3 (good health and well-being), 10 (reduced inequalities), 11 (sustainable cities and communities), and 13 (climate action)
Decarbonization	Goals 3 (good health and well-being), 7 (affordable and clean energy), 13 (climate action), 15 (life on land), and 17 (partnerships for the goals)
Science, Technology, and Innovation	Goals 9 (industry, innovation, and infrastructure), 12 (responsible consumption and production), and 17 (partnerships for the goals)
Science and Peace	Goals 5 (gender equality), 10 (reduced inequalities), 16 (peace, justice, and strong institutions), and 17 (partnerships for the goals)
Financing to Achieve the SDGs	Goals 5 (gender equality), 9 (industry, innovation, and infrastructure), 10 (reduced inequalities), and 17 (partnerships for the goals)

chose these themes for its information-gathering purposes due to their diverse characteristics in terms of varied sustainability challenges, such as environmental, economic, and social considerations. Although each theme affects almost all of the SDGs in important aspects, key relevant SDGs are listed for each.

The remainder of this report is organized by workshop theme (Chapters 2–9). Several caveats should be kept in mind, however. First, just as the SDGs are interrelated, the challenges and solutions discussed at each workshop overlap. For example, issues related to food systems emerged during discussions about financing; education was a factor in presentations about urbanization; and peace and conflict were identified as critical conditions to understanding food systems. Second, in keeping with the Statement of Task, the committee did not undertake a full consideration of every issue embedded in every SDG. The challenges and case studies in the following chapters—and in callout boxes—reflect those highlighted by the workshop presenters. The committee drew on the workshop discussions, previous National Academies reports and workshops, a bounded literature review, and committee members' expertise to develop key research priorities and possible actionable steps related to each theme. The workshops were the primary source of information from which the key research priorities and potential actionable steps were developed. The final chapter (Chapter 10) compiles and organizes the committee's key research priorities and possible actionable steps by stakeholder. The committee believes that these recommendations are ambitious but realistic and, taken together, can make measurable progress toward a sustainable future for all.

2

Education and Capacity Building

The goal of a quality education is articulated through Sustainable Development Goal (SDG) 4 ("Ensure inclusive and equitable quality education and promote lifelong learning opportunities for all."). Education is critical to achieving the other SDGs, and educational institutions at all levels are powerfully positioned to operationalize sustainability across society. In addition, educational institutions have an important role to play in educating about the SDGs themselves.

CHALLENGES

Within the United States—whether within the federal government, media, educational systems, or other domains—public knowledge about the 2030 Agenda is limited, in terms of actions being taken both globally and domestically that align with the SDGs (Mendelson, 2022; World Economic Forum, 2019). Among the 193 United Nations (UN) member states, 187 countries have undertaken Voluntary National Reviews (VNRs), which provide an intentional way to identify gaps and chart future action (UN, 2022e). The 6 countries that have not submitted VNRs are Haiti, Iran, Myanmar, South Sudan, Yemen, and the United States (Sachs et al., 2022).[1] When the SDGs are recognized in the United States, many people assume they are focused on environmental improvement and/or are intended for the Global South (Mendelson, 2022b).

Building sustainable mindsets begins at a young age, but there is little effort at the K-12 and university levels to apply inquiry-based learning to help students

---

[1] In 2022, Brookings and the UN Foundation released *The State of Sustainable Development Goals in the United States* as a shadow VNR, calling for stronger U.S. leadership on the SDGs (Pipa et al., 2022).

learn about the SDGs and to develop partnerships of stakeholders to understand local needs and develop actionable steps toward progress. As one presenter posed, "What are we doing to educate our students to understand the complex global challenges of our time?" (O'Donnell, 2022). She pointed to SDG Target 4.7, which states that "by 2030, ensure all students acquire knowledge and skills needed to promote sustainable development." Achievement of this goal requires making complex subjects understandable, building mindsets for long-term engagement, changing abstract SDGs into locally relevant issues, and taking action for change—and engaging children at a young age.

CASE STUDIES AND SYNERGIES

The committee learned of promising initiatives that address these challenges in creative and implementable ways. Presenters described experiential learning and community partnerships that advance all the SDGs, while providing the "quality education" articulated in SDG 4. At the K-12 level, for example, the Smithsonian Science Education Center has created opportunities for students to learn about local issues to build global sustainability mindsets (Box 2-1; Figure 2-1). At the university level, Carnegie Mellon University (CMU) launched a campus-wide, multidisciplinary Sustainability Initiative that resulted in, among other things, the first Voluntary University Review (VUR) to assess how education, research, and practice in a postsecondary educational setting align with the SDGs (Box 2-2). CMU has also supported the city of Pittsburgh in conducting a Voluntary Local Review and has involved students in capstone research and action projects in Pittsburgh and other cities.

Other examples discussed include a "Sustainability 101" course that all students at Arizona State University, regardless of major, will take; regional partnerships spearheaded by the University of Texas at Arlington to promote Education for Sustainable Development (Tare, 2022); student-led SDG projects in partnership with the City of Los Angeles (Apolitical, 2022a); and transformation of the university campus as a model for sustainability practices for the surrounding community, as done at Chiang Mai Rajabhat University in Thailand. Drawing on the concept of "glocal" or meso scale of communities, Marc McCaffrey (The Long Game) noted the multiplying impact of the "Powers of 10": that is, the networks that an individual can tap into to create change over time (McCaffrey, 2022).

A number of studies examine key competencies for sustainability in higher education. Wiek et al. (2011) discuss five core competencies in sustainability education: a systems thinking competence, an anticipatory competence, a normative competence, a strategic competence, and an interpersonal competence. A Delphi study with 14 international experts in sustainability education extends this framework to propose two additional key sustainability competencies: an implementation competency and an intrapersonal competency or mindset (Brundiers et al., 2020).

BOX 2-1
K-12 Education: Smithsonian Science for Global Goals

Experiential learning in which students undertake local investigations that connect to global issues is an effective way to educate and inspire young people. One example is the work of Smithsonian Science Education Center (SSEC), through the Smithsonian Science for Global Goals project.

Mission and Goals

The mission of SSEC is transforming K-12 Education Through Science™ in collaboration with communities across the globe. SSEC is the only organization within the Smithsonian Institution that focuses specifically on formal K-12 STEM (science, technology, engineering, and math) education, according to its director, Carol O'Donnell. To achieve its interlocking goals of innovation, sustainability, and inclusion, SSEC promotes active inquiry-based STEM teaching and learning; advances K-12 STEM education for sustainable development; and ensures diversity, equity, accessibility, and inclusion in K-12 STEM education. SSEC involvement with the InterAcademy Partnership, an umbrella group of more than 140 science and medicine academies, supports global implementation.

Developing Sustainability Mindsets

In 2016, SSEC intentionally aligned its work with the United Nations Sustainable Development Goals (SDGs), creating the Smithsonian Science for Global Goals project in collaboration with the InterAcademy Partnership to help young people "discover, understand, [and] act." Locally relevant, locally driven, but globally important experiential learning experiences combine STEM education, social and emotional learning, and civic engagement.

SSEC has developed a set of *Community Research Guides* on four multifaceted subject areas: sustainable communities, food, mosquitos, and biodiversity, with each curriculum expected to take about 16 weeks to complete. For example, the Biodiversity Guide is a seven-part curriculum that helps students answer a vital question: How can we balance the needs of people with the needs of other living things? By working through research and other activities in the Food Guide, students develop a community action plan to ensure good nutrition for all. The Mosquito Guide helps students in the United States, Africa, and Central America develop different sets of actions based on the problems they identify. Shorter *Community Response Guides* focus on more specific issues—currently, COVID-19, vaccines in general, and environmental justice. More guides are being developed—focusing on climate change, biotechnology, extreme weather, and more—and will align with the SDGs.

References

IAP (The InterAcademy Partnership) website, https://www.interacademies.org/iap/about.
O'Donnell, C. 2018. Science Education, Identity, and Civic Engagement: Empowering Youth through the UN Sustainable Development Goals. G7 Executive Talk Series.
O'Donnell, C. 2022. Workshop Presentation, April 18, 2022.
Pahnke, J., C. O'Donnell, and M. Bascope. 2019. Using Science to Do Social Good: STEM Education for Sustainable Development. Position paper at the second International Dialogue on STEM Education, Berlin, Germany.
SSEC website, https://ssec.si.edu.

FIGURE 2-1 Building sustainability mindsets.
SOURCE: Carol O'Donnell, Workshop Presentation, April 18, 2022.

KEY RESEARCH PRIORITIES FOR EDUCATION
AND CAPACITY BUILDING

To operationalize sustainable development in areas relating to education and capacity building, the field could prioritize the following research activities:

- Conduct research investigations, case studies, and evaluations of effective efforts building partnerships and operationalizing the SDGs at the local and subnational levels that connect to national and global levels with special focus on K-12 and university education, public outreach, and capacity building.
- Identify effective ways to support K-12 education initiatives that assist students with defining, developing, and implementing their own frameworks for sustainable actions in their communities and in understanding the impacts beyond.
- Examine issues relating to ensuring diversity, equity, accessibility, and inclusion in K-12 STEM education as well as leveling the playing field in access to K-12 education across school districts in the United States.
- Examine how sustainability education programs at the undergraduate and graduate levels can prepare all students, regardless of major, to contribute to advancing a post-2030 agenda for sustainable development, as well as identify best practices in field building for sustainable development at the undergraduate and graduate levels that will be important for research and education in moving that agenda forward.

BOX 2-2
Higher Education: Carnegie Mellon University

Carnegie Mellon University (CMU) serves as an example of how colleges and universities can operationalize sustainability—internally and externally.

CMU's Sustainability Initiative

CMU has more than 14,000 students in its seven schools and colleges. The university launched a Sustainability Initiative in 2019, a renewal of sustainability activities in education, research, and practice that began in the 1990s. At that time, a Green Design education and research initiative and a Green Practices Committee were established. CMU has been active with the Association for the Advancement of Sustainability in Higher Education (AASHE) since it was founded in the early 2000s. Internal studies in recent years, including the 2017 "Pittsburgh to Paris: Reducing the Carbon Footprint of Carnegie Mellon University" and a 2019 Task Force on Campus Climate, prompted the Sustainability Initiative. Established under the leadership of the provost, the initiative made a commitment to "help solve pressing problems brought to light by the SDG framework, by acting boldly, taking risks, and applying creativity." Another commitment was to complete a Voluntary University Review (VUR) by September 2020.

Voluntary University Reviews: 2020 and 2021

CMU's VUR is the first university review in the world to map to the Sustainable Development Goals (SDGs). According to Provost James H. Garrett, in his preface to the 2020 VUR, "Our intention is for CMU's VUR to be a framework for us to track what we are doing across the 17 Global Goals and where we might find opportunities to do more." As a baseline, a survey was undertaken in February 2020 to take stock of the CMU community's understanding of and interest in the SDGs. Most respondents were not aware of them but expressed interest in learning more. Hundreds of students, faculty, and staff engaged in the process. A steering committee appointed by the provost includes senior faculty members (workshop presenter Sarah Mendelson is one), the dean of libraries, a staff director of the initiative, and the assistant vice president for facilities. The VURs, conducted with the aid of data analytics tools, look across CMU's education, research, and practice missions. Examples of findings are:

- Education—Reviewed the 2,938 courses offered in spring 2020 to determine whether/which SDGs were addressed;
- Research—Reviewed 995 CMU publications between 2018 and 2020 to identify sustainability-related topics;
- Practice—Reviewed reports, support centers, policies, and student organizations; operations, such as facilities management, housing, and transportation services; and human resources.

A 17 Rooms exercise, modeled after the innovative 17 Rooms Initiative of the Brookings Institution and The Rockefeller Foundation, was conducted in May 2020 (by videoconference due to the pandemic) to engage mixed groups of students, faculty, and staff with interests in particular SDGs and how they are or could be addressed at CMU. Through the 17 Rooms Initiative, stakeholders come together virtually to identify actionable steps around each of the SDGs that "are big enough

BOX 2-2 Continued

to matter and small enough to get done" over the period of 12 to 18 months. During summer 2020, seven students reviewed, synthesized, interpreted, and followed up on the initial set of findings from the CMU 17 Rooms. They mapped each activity to an SDG, recognizing some subjectivity and the relationship of many activities to more than one SDG. These analyses helped support the 2020 VUR. In 2021, the VUR focused on innovations in collection, analysis, and sharing of information. An algorithm (open source for others to use) for analysis of CMU course offerings was developed to enable analysis of many more courses. Similarly, research and practice activities can be more extensively examined through interactive data analytics tools.

Other Efforts
    With the release of the VUR, outreach included participation in a side event at the September 2019 United Nations (UN) General Assembly hosted by Brookings and the UN Foundation and engagement with various higher education networks. CMU students have also been involved in creating a Voluntary Local Review for the city of Pittsburgh. Mendelson described a student capstone project to develop case studies in several U.S. and Canadian cities tracking the impact of pandemic relief and recovery funds on issues relating to social justice needs (including SDGs 2, 3, 10, and 16). She is also editing a volume of case studies that examine teaching human rights differently using the SDGs.

References

    17 Rooms website, https://www.brookings.edu/project/17-rooms.
    CMU (Carnegie Mellon University). 2022. The Sustainability Initiative at CMU: Toward a more sustainable future—for everyone. https://www.cmu.edu/leadership/the-provost/provost-priorities/sustainability-initiative/index.html.
    CMU. 2021. *CMU Voluntary University Review of the Sustainability Development Goals 2021.* https://local2030.org/library/848/Carnegie-Mellon-University-Voluntary-University-Review-2021.pdf.
    CMU. 2020. *CMU Voluntary University Review of the Sustainability Development Goals 2020.* https://www.cmu.edu/leadership/the-provost/provost-priorities/sustainability-initiative/cmu-vur-2020.pdf.
    Mendelson, S. 2022a. Heinz in the Hague. https://www.heinz.cmu.edu/media/2022/July/heinz-in-the-hague.
    Mendelson, S. 2022b. Workshop Presentation, April 18, 2022.

POSSIBLE ACTIONABLE STEPS FOR
EDUCATION AND CAPACITY BUILDING

Possible actionable steps for undertaking inquiry-based education initiatives and capacity-building partnerships essential to making progress on the SDGs at the local, national, and global levels are as follows:

- University leaders could undertake initiatives to assist faculty and students in developing VURs to evaluate needs and prioritization among

SDGs based on an institutional mission, take actionable steps that advance progress on the SDGs at their universities, and ensure that every student regardless of major is exposed to the challenges and opportunities in sustainable development.

- University leaders and faculties could develop partnerships with local and national governments, universities, business communities, and civil society organizations to develop VURs to evaluate needs and take actionable steps that can advance progress toward the SDGs by their cities and local communities (Apolitical, 2022b).

- University leaders and faculties could elevate a focus on building the field of sustainability science as a discipline to prepare the next generation for a post-2030 agenda for sustainable development.

- Cities and school districts could initiate and support programs at the local K-12 level for students to undertake local investigations in their communities on the SDGs across diverse contexts; define and implement frameworks for sustainable development; and connect their local issues to global issues, such as the Smithsonian Science for Global Goals (O'Donnell, 2022).

- U.S. government and education leaders could engage the public to raise awareness of the SDGs. Examples include the National Climate Assemblies in many European countries, such as Austria, Denmark, Germany, Finland, France, Scotland, Spain, and the United Kingdom, that discuss the climate crisis with randomly selected citizens (Bürgerrat, 2022); global campaigns such as the World's Largest Lesson (2022) and World's To Do List (2022); São Paulo's Municipal Agenda 2030 and its public policy councils (Open Government Partnership, 2022); and work with storytellers, such as Sony's "Picture This" short film competition.

- Urban leaders and planners could engage students from universities, community colleges, and Minority-Serving Institutions (NASEM, 2020) to organize student projects in local communities and cities, such as the student-led research to help develop the City of Los Angeles' Biodiversity Index (City of Los Angeles, 2022).

- Education leaders could provide teachers with peer mentor networks and a platform, such as developing and maintaining a website to host downloadable materials relating to sustainability, the SDGs, and climate change education.

3

Localization of the Sustainable Development Goals and Indigenous Knowledge

The Sustainable Development Goals (SDGs) embrace global aspirations, but they must be rooted in local buy-in and implementation. According to the United Nations (UN) Global Task Force on Local and Regional Governments, "localization relates both to how the SDGs can provide a framework for local development policy and to how local and regional governments can support the achievement of the SDGs through acting from the bottom up" (UN, 2016). Local communities have a role to play in achieving the SDGs. Further, "while the SDGs are global, their achievement will depend on our ability to make them a reality in our cities and regions" (UN, 2016). Two interrelated solutions emerged from the workshops: localization of the SDGs to municipalities and sub-regions and rural villages/communities, and recognition of the value of indigenous knowledge for sustainable development.

CHALLENGES

Many of the challenges to operationalizing the SDGs at the local and national levels involve people. Local officials and leaders are in a central position to implement the SDGs because of the range of services they must provide, but their large menu of responsibilities is rarely accompanied by sufficient budget and other resources. Although some of the most effective sustainability leaders are embedded in local settings and view the SDGs as complementary to their goals, others perceive the SDGs as yet another competing priority. Even when embraced, scarce resources limit the capacity to follow best practices or maintain ongoing SDG initiatives. For example, among the 100 largest cities in the United States, 45 have pledged to reduce their greenhouse gas emissions from a baseline level. But only 32 conducted at least one follow-up from the baseline since 2010, and only 26 actually experienced a decrease compared to their baseline levels (Pipa, 2022).

Capacity challenges exist everywhere at the subnational level. Elected leaders subject to the political cycle face another set of pressures and competing priorities. Even when a locality wants to do the right thing, jurisdictional boundaries, regulatory limitations, and financing can create obstacles. A lack of data standards and disaggregated data collection at the local level further cloud the picture. Disaggregated data are critical to understand trends across different population groups and to ensure adherence to the core element of the SDGs—"Leave No One Behind" (UN, 2022c).

Indigenous knowledge has enabled many communities to sustain themselves for hundreds, if not thousands, of years, for example related to food production, water management, and medicine. Some authors claim that indigenous philosophers who criticized European customs inspired European intellectuals such as Jean Jacques Rousseau and initiated the European Enlightenment (Graeber and Wengrow, 2022). Yet indigenous knowledge communities are often marginalized socially and economically, and their knowledge viewed as "un-scientific." Conversely, when indigenous knowledge is tapped, care must be taken to develop mutually beneficial relationships and to not simply appropriate what indigenous knowledge offers to achieve the SDGs (IFAD, 2019; Jessen et al., 2021).

CASE STUDIES AND SYNERGIES

Localization calls for the political will to connect the SDGs to the role that local governments play in improving the lives and livelihoods of citizens. The emphasis on localization does not mean that cities should go it alone. In addition to the SDG Leadership Cities Network and Local 2030 Hubs mentioned in Chapter 4, resources include the African Network of Cities (Parnell, 2022) and United Cities and Local Governments (UCLG), with 250,000 members (Saiz, 2022). An important UCLG activity is sharing successful examples of localization of the SDGs, as well as training materials for practitioners.

The committee's workshop session on localization underscored the value of Voluntary Local Reviews (VLRs). UCLG has found that when these reviews occur, the dialogue between and across levels of government has been positively influenced to accelerate progress toward the SDGs. In addition, discussion about service delivery provision shifts. Going deeper than the national level can reveal where new investments and indicators are necessary, and can help break down internal silos (Pipa, 2022). As shown in Pittsburgh, development of a VLR can also connect local academic institutions to the surrounding community (Mendelson, 2022a). The UN Foundation provides examples of implementation of VLRs in U.S. cities, such as New York City, Hawaii, Los Angeles, Orlando, and Pittsburgh (UN Foundation, 2022). The Swedish International Centre for Local Democracy (ICLD) describes examples in cities across the globe, including successful practices employed in small towns such as Shimokawa, Japan, with a population of roughly 3,000 (Ciambra and Martinez, 2022).

Local implementation of the SDGs has led to such innovations as multi-stakeholder boards in Bristol, England, and more integrated regional planning in Orlando, Florida (Pipa, 2022). As noted previously, electoral cycles can be tricky, especially when the SDGs, or sustainability in general, are tied too closely to a leader no longer in office. Durability has taken different forms, from citizen engagement in Mannheim, Germany, to master planning in Bogota, Colombia. Embedding sustainability into government, citizens groups, and informal networks provides the longevity needed

As an on-the-ground example of localization, Hawai'i Green Growth uses and contributes knowledge through a number of networks (Box 3-1). The common language of the SDGs can be used to produce data that are transparent and useful.

BOX 3-1
Hawai'i Green Growth

Sustainability on the islands of Hawai'i goes back 1,000 years before European contact. Hawai'i Green Growth (HGG) draws from the lessons of this "original circular economy," according to Celeste Connors, HGG executive director.

Economic Recovery through Sustainable Development

HGG brings together more than 150 diverse stakeholders committed to economic, social, and environmental priorities. It was launched during the 2011 Asia Pacific Economic Cooperation Summit, which was held in Honolulu to identify green growth opportunities and advance a green economy within an island context. According to Connors, "green" is in the organization's name, but it is inherently an economic recovery strategy that grew from the 2008 global financial crisis and remains relevant today. Success rests on four key pillars: leadership commitment, public-private partnerships, measurement of progress, and concrete, on-the-ground action. Hawai'i's statewide sustainability commitment is embodied in the Aloha+ Challenge in the six priority areas of natural resources management, local food, smart sustainable communities, green workforce and education, waste reduction, and clean energy. It was recognized by the United Nations as one of the first Local2030 sustainability hubs. The hubs are recognized as having locally appropriate responses to sustainable development even prior to adoption of the Sustainable Development Goals (SDGs).

Priorities and Data Sharing

Five working groups have been formed that work across the priority areas. An online open data dashboard built on an ESRI-based platform tracks accountability and actions related to these areas. It draws from government, business, and community-driven data. The next phase will look at scenarios to project the impact and tradeoffs and the further engagement of youth, fostering their bottom-up knowledge of the SDGs to position them to lead. Hawai'i was also the first U.S. state to conduct a Voluntary State Review in 2020.

BOX 3-1 Continued

Local2030 Islands Network

Building on what has happened over the past decade, HGG co-leads the Local2030 Islands Network, a peer-to-peer network of national, state, and local economies committed to advancing the SDGs through locally driven solutions. At the 2020 Our Ocean Conference in Palau, the network announced a series of commitments with ESRI to support island SDG Dashboards.

References

Connors, C. 2022. Workshop Presentation, April 21, 2022.
Hawai'i Green Growth website, https://www.hawaiigreengrowth.org.
Local2030 Islands Network website, https://www.islands2030.org.

A public data dashboard shows citizens the progress being made and where work is falling short. Measuring what matters encourages multistakeholder-driven development of local metrics and indicators and an understanding of how diverse metrics are related to each other and to the SDGs. "Progress moves at the pace of trust," Celeste Connors stressed during one workshop session (Connors, 2022). She also called attention to a vital SDG that is often overlooked—SDG 17, Partnership. People often want to take quick action, but process matters; it takes time to convene and connect diverse partners, identify shared priorities, measure what matters, and

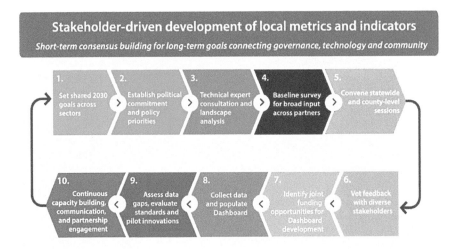

FIGURE 3-1 Stakeholder-driven development of local metrics and indicators.
SOURCE: Celeste Connors, Workshop Presentation, April 21, 2022.

coordinate to drive action. To measure what matters, it is necessary to determine the values and priorities of different stakeholders, which Hawai'i Green Growth does with an iterative 10-step process across locations and sectors (Figure 3-1).

Similarly, while indigenous communities have traditionally looked inward to develop solutions, such as in India (Box 3-2), the Global Tapestry of Alternatives connects indigenous and other "alternative" communities around the world.

BOX 3-2
New Paradigms That Draw from Indigenous Knowledge

Dealing with contradictions and tradeoffs inherent within the Sustainable Development Goals (SDGs) is at the heart of operationalizing sustainability. Ashish Kothari, a founder-member of the nonprofit Kalpavriksh, offered several examples from India, as well as a new paradigm for sustainable transformations. There are thousands of examples of people around the world trying to meet their needs and aspirations while harmonizing with the rest of nature and creating equality and justice, Kothari said. In India, his group has documented or collected 1,800 examples of people on the ground with alternative initiatives for well-being to meet daily needs, address social issues, and strengthen local governance. A few examples follow:

- **Community sovereignty and social justice:** In southern India, the Deccan Development Society has empowered 5,000 Dalit women to change farming practices through seed diversity shared in community grain banks; rain-fed, organic agriculture; and land reform. They have achieved household food security and have control over the food system, that is, food sovereignty.
- **Community self-mobilization:** In Kunariya (Kachchh), a community in western India with multiple castes, the local panchayat (government) has taken action toward health and relative economic self-reliance. It stood the test when the village was able to withstand COVID-19 while continuing its economic activity.
- **Multi-community federation for self-governance:** In Gadchiroli (Maharashtra, central India), 90 villages have formed a Maha Gram Sabha, a federation of village assemblies, to combat mining proposals, conserve forests, and generate sustainable livelihoods. At the same time, women's empowerment and cultural identity are strengthened.
- **Urban community-led planning:** In northwestern India, in Bhuj town, Kachchh, economically marginalized urban dwellers did their own planning (with facilitation by five civil society organizations in a program called Homes in the City) to re-imagine common spaces; develop water self-sufficiency, waste management, and sanitation solutions; and empower women and girls to be part of decision-making. They assert that urban planning must be decentralized to the local level.

Eco-Swaraj
Kothari offered a way to reimagine the future with transformations in five spheres: ecological resilience and wisdom, cultural and knowledge diversity, radical democracy,

BOX 3-2 Continued

economic democracy, and social justice and well-being. At the core is a shift of values and principles, which he said must be at the fulcrum of well-being, going beyond current notions of sustainable development. For example, lessons from grassroots initiatives show that cooperation is more important than competition, the commons are more important than private property, and solidarity makes for better coping. One framework encompassing this is "eco-swaraj" or Radical Ecological Democracy, which posits that everyone is central to decision-making and that the principles of equity, justice, and sustainability (including respect of the rest of nature) could guide decision-making. Similar worldviews and frameworks exist across the world, in ancient indigenous cosmologies, radical reinterpretations of religions, and counter-currents within the industrialized west.

In India, Vikalp Sangams (Alternative Confluences) bring together different movements for collective visioning and collaboration. Worldwide, the Global Tapestry of Alternatives provides these opportunities to create a critical mass for change.

References

Global Tapestry of Alternatives website, https://globaltapestryofalternatives.org.
Kothari, A. 2022. Workshop Presentation, April 29, 2022.
Kothari, A., A. Salleh, A. Escobar, F. Demaria, and A. Acosta. 2019. Pluriverse: A Post-Development Dictionary. Tulika Books and Authors Upfront, Delhi.
Radical Ecological Democracy website, www.radicalecologicaldemocracy.org.
Vikalp Sangam website, www.vilkapsangam.org.

KEY RESEARCH PRIORITIES FOR LOCALIZATION OF THE SUSTAINABLE DEVELOPMENT GOALS

The committee proposes the following key research priorities to operation-alize sustainable development to localize the SDGs, building on a synthesis of research gaps developed by the International Science Council (ISC, 2021):

- Understand the synergies and tradeoffs that can help to achieve localization of the SDGs, including the appropriate balance between economic, social, and environmental considerations at the local, national, and global levels.
- Identify key mechanisms that address poverty and empower vulnerable communities.
- Identify governance models and arrangements that could accelerate local transformations for sustainable development.
- Explore ways to make science systems more inclusive and equitable, to involve a wider range of voices, institutions, types of knowledge, and approaches to learning that are designed to capture local needs.
- Establish effective frameworks that incorporate both conventional scientific knowledge and indigenous knowledge.

POSSIBLE ACTIONABLE STEPS FOR LOCALIZATION OF THE SUSTAINABLE DEVELOPMENT GOALS

The committee identifies the following possible actionable steps to operationalize sustainable development to localize the SDGs:

- The U.S. government could commit to creating a Voluntary National Review (VNR) by encouraging more states and cities to conduct VLRs and synthesize already good work at the local level to scale to a VNR roll-up.
- Local officials could commit their support to the SDGs and use the framework to align local policies and initiatives.
- Urban and community leaders and planners as well as philanthropic organizations could learn from excellent case studies of knowledge networks, such as C40 (2022), the Brookings' City Playbook for Advancing the SDGs (2021), Global Islands Partnership (2022), UCLG Learning (2022), Vikalp Sangam (2022), Global Tapestry of Alternatives (2022), and African Network of Cities (2022), which effectively incorporate indigenous knowledge for advancing sustainability.
- U.S. universities could help surrounding communities and cities conduct Voluntary University Reviews (VURs) and/or VLRs (as described in Chapter 2 on Education and Capacity Building).
- The federal government could provide financial incentives for local and state VLRs and consider federal and state regulatory changes to create flexibility. A scalable model would include Hawai'i's Open-Data Aloha+ Challenge process and Dashboard (2022).
- Funding agencies and philanthropic organizations could support initiatives that further the role of indigenous knowledge in the development of scientific knowledge. Examples include the National Center for Complementary and Alternative Medicine of the National Institutes of Health (2011), the Beijing Center (2022) focusing on the introduction of traditional medicine in China, and efforts to codify traditional medicine in health care and services (Mashelkar, 2001).

4

Food Systems

The food system's impact on daily life is widespread. Food production encompasses the supply of agricultural inputs; crop, livestock, fish, and commodities; storage, transportation, processing, and packaging; retailing; and preparation. On the consumption side, whether individuals eat nutritiously or are hungry or malnourished, whether they are healthy, obese, or underweight, whether they live in urban or rural areas, they are involved in, and greatly affected by, the food system. Although Sustainable Development Goal (SDG) 2 speaks to ending hunger, food systems impact all 17 SDGs. Achieving the SDGs will be impossible without a holistic reform of the food system, taking into account environmental, economic, and social considerations.

CHALLENGES

The food system, starting with agricultural production, has large-scale environmental impacts. It accounts for one-third of global greenhouse gas emissions (Crippa et al., 2021), while irrigation for agriculture accounts for approximately 70 percent of global water use (World Bank, 2022b). Agriculture is also responsible for 80 percent of changes that threaten biodiversity and other planetary resources and processes (Müller, 2022).

Despite improved agricultural techniques, between 720 and 811 million people in the world went hungry and approximately 2.4 billion people (30 percent of the global population) lacked access to adequately nutritious food in 2020, even before the current rise in global food prices (UN, 2021c) and supply chain disruption. This population includes 38.3 million people in the United States (USDA, 2021). Tragically, an abundance of food alongside a high incidence of food insecurity is also accompanied by an abundance of food loss and waste (NASEM, 2019). The U.S. Department of Agriculture (USDA) estimates that 30 to 40 percent of U.S. food is lost or wasted each year (USDA, 2022), and the Ellen MacArthur Foundation produced similar worldwide estimates (Ellen MacArthur Foundation, 2022). Food loss and waste occur throughout the supply chain, with most loss occurring in the early and middle stages of the food chain in developing countries (on-farm activities, transport, storage, processing, and wholesaling) and significant waste occurring in the consumption stage in developed countries (UNDP, 2022c). Given the realities of geopolitics and supply and demand, reducing food waste is unfortunately not an automatic solution to food insecurity and urban hunger (Grantham, 2022).

Food-system policies and programs have complex interactions with equity. Cities with high poverty rates and high costs of living generally have the highest food insecurity, and such factors as population density, household size, and income affect the incidence of this insecurity (Grantham, 2022). Poverty and equity play out in access to nutritious food, with implications for human health. In Washington, DC, for example, access to grocery stores is extremely limited in lower-income neighborhoods, whose populations are predominantly non-Hispanic Black. These neighborhoods also have the highest percentage of adult residents who are obese and/or have diabetes (Figure 4-1).

Also important is consideration of how new changes to food systems affect culture, affordability, and the livelihoods of those who practice the "business as usual" (Kanbur, 2022). As shown through work by the Food Systems Economic Commission, transformation creates winners and losers, and distributional consequences of such need to be addressed.[1] For example, reducing beef consumption is promoted for health and environmental benefits, but millions of people worldwide rely on cattle for their livelihoods. In terms of food security, it is essential to recognize the importance of availability, access, diversity, and stability for a food system that serves the needs of diverse stakeholders.

CASE STUDIES AND SYNERGIES

During the workshop, presenters suggested new approaches to food production, including regenerative (Box 4-1) and restorative urban (Box 4-2) agriculture. Regenerative agriculture involves "a system of farming principles that rehabilitates the entire ecosystem and enhances natural resources, rather than depleting them" (Rodale Institute, 2020). Restorative urban agriculture involves investing in decentralized, neighborhood-based, controlled environment agriculture that uses water, heat, and light in energy-efficient ways and brings sustainable, regenerative food production closer to poor urban neighborhoods that currently lack affordable access to healthy foods (O'Hara, 2022).

---

[1] See https://www.foodsystemeconomics.org/about-the-commision.

FIGURE 4-1 Inequitable distribution of full-service grocery stores in Washington, DC.
NOTE: Four Urban Food Hubs are established in the food desert neighborhoods to enhance food security (see Box 2-2).
SOURCE: Sabine O'Hara, Workshop Presentation, April 18, 2022, and O'Hara, 2017.

BOX 4-1
Regenerative Agriculture: Regenerate Costa Rica

Regenerative agriculture promotes interdisciplinary approaches to land use management through nature-based solutions. Regenerative agriculture can capture large amounts of carbon in the soil, reverse land degradation, reduce the use of fertilizers, retain water, and mitigate climate change. The concept is based on regenerative development, which holistically integrates six fundamental pillars: Mother Earth, Society, Economy, Spirituality, Politics, and Culture (Müller, 2022). Regenerative agriculture promotes resilience and community by converting dry soils to highly productive ecosystems, providing educational empowerment for

BOX 4-1 Continued

children and communities, producing abundant and healthy food, and incorporating indigenous knowledge (Müller, 2022).

Through Regenerate Costa Rica, community farms in Guanacaste, in northwestern Costa Rica, promote regenerative agriculture, including the recovery of biodiversity, holistic livestock, and local education connected to the biological, geographical, and cultural environment. In the central part of the country, Finca Rosa Blanca, a coffee farm and eco hotel, promotes regenerative agriculture with regenerative tourism (Essential Costa Rica, 2022).

The carbon sequestration potential of global adoption of regenerative agriculture has been studied (Rodale Institute, 2020; see also Chapter 6), and data, information, and knowledge need to be turned into wisdom and action to protect the planet. The Regenerate Costa Rica initiative is creating a national network to share skills, knowledge, and solutions in collaboration with the Regenerative Communities Network. Müller calls for holistic approaches using scientific, indigenous, and local knowledge; research that enables immediate action; actions conducive to all life for regeneration; local well-being above corporate and trade; and education for young people with hope and guidance.

References

Essential Costa Rica. 2022. Costa Rica Takes Sustainable Travel to the Next Level. https://www.visitcostarica.com/en/costa-rica/blog/costa-rica-takes-sustainable-travel-next-level#:~:text=Finca%20Rosa%20Blanca&text=As%20one%20of%20the%20original,directly%20benefit%20the%20surrounding%20community.

Müller, E. 2022. Workshop Presentation, April 18, 2022.

Rodale Institute. 2020. Regenerative Organic Agriculture and the Soil Carbon Solution. https://rodaleinstitute.org/education/resources/regenerative-agriculture-and-the-soil-carbon-solution.

BOX 4-2
Restorative Urban Agriculture:
The Food Hubs Initiative in Washington, DC

One of the urban agriculture initiatives of the University of the District of Columbia (UDC), the only public university in the nation's capital, and the only exclusively urban land-grant university in the United States, is a circular economy approach to addressing food insecurity in food desert neighborhoods in Washington, DC (O'Hara, 2017, 2022). The U.S. Department of Agriculture defines food deserts as "urban neighborhoods and rural towns without ready access to fresh, healthy, and affordable food" (USDA, 2015). Food access is uneven

BOX 4-2 Continued

among Washington, DC's eight administrative Wards. Most of the city's food deserts are located in Wards 7 and 8, east of the Anacostia River, which forms a dividing line in the city. Neighborhoods with the lowest household incomes, the highest unemployment rates, and the largest percentage of non-Hispanic Black residents are located east of the river. Food insecurity and food-related health problems such as obesity and diabetes are also significantly higher in those neighborhoods (O'Hara, 2018; O'Hara and Toussaint, 2021).

In response, UDC developed an Urban Food Hubs model that consists of four integrated components and illustrates a circular economy system: (1) food production, (2) food preparation, (3) food distribution, and (4) waste and water management. Food production occurs in small spaces including rooftops, parking lots, and between buildings, and utilizes innovative hydroponics and aquaponics systems. Food preparation and nutrition education are offered in community kitchens and food trucks. Food distribution engages farmers markets, community-supported agriculture, restaurants, and small stores. Waste and water management activities include composting, water harvesting, rain gardens, and other green infrastructure initiatives (O'Hara, 2017). The model is scalable and can form the core of a larger, regional food-based economy (Stuiver and O'Hara, 2021). Four Urban Food Hubs have been built to date, three of them in food desert neighborhoods in DC's Wards 5, 7, and 8 (Figure 2-1).

The Urban Food Hubs serve as teaching and innovation hubs to provide local communities with sustainable alternatives to food access while addressing economic development and green infrastructure needs. They represent a holistic view of a food system that restores both productive and absorptive capacities of the environment through a decentralized approach that offers economic opportunities, improves public health, alleviates disparities, and produces food where the majority of consumers live, namely in cities and metro areas. By scaling them, the Urban Food Hubs can lower the energy footprint of food, reduce water use and food waste, and reconnect people with nature (O'Hara, 2022).

References

O'Hara, S. 2022. Workshop Presentation, April 18, 2022.

O'Hara, S. 2018. The Five Pillars of Economic Development: A Study of a Sustainable Future for Ward 7 and 8 in Washington, D.C. https://docs.udc.edu/causes/Five-Pillars-DC-Final-05-2018.pdf.

O'Hara, S. 2017. The Urban Food Hubs solution: Building capacity in urban communities. *Metropolitan Universities* 28(1).

O'Hara, S., and E. C. Toussaint. 2021. Food access in crisis: Food security and COVID-19. *Ecological Economics* 180:106859. https://doi.org/10.1016/j.ecolecon.2020.106859.

Stuiver, M., and S. O'Hara. Food connects Washington DC in 2050—A vision for urban food systems as the centerpieces of a circular economy. *Sustainability* 13(14):7821. https://doi.org/10.3390/su13147821.

USDA (U.S. Department of Agriculture). 2015. NAL Agricultural Thesaurus. https://agclass.nal.usda.gov/vocabularies/nalt/concept?uri=https://lod.nal.usda.gov/nalt/142350.

Discussion of sustainable food systems occurred throughout the committee's workshop series. For example, during the session on decarbonization (see Chapter 6), biochar was presented as having the potential to improve soil fertility and crop productivity and mitigate global warming by efficiently sequestering large amounts of carbon in soil over the long term, as well as to improve yield and reduce the need for chemical fertilizers (Draper, 2022; Semida et al., 2019). New York City's efforts to increase local food supply have included establishment of the city's first Office of Urban Agriculture and plans for the private sector to convert Riker's Island from a prison to an urban farm serving predominantly poorer New York City neighborhoods (Koval, 2022).

KEY RESEARCH PRIORITIES FOR FOOD SYSTEMS

The concept of food systems is much broader than agriculture. Based on the presentations and discussions during the April 2022 workshop and its literature review, the committee identified key priorities for research in order to operationalize sustainable development in the area of sustainable and equitable food systems:

- Conduct a comprehensive analysis of the entire food system, including production, processing, distribution, consumption, and waste disposal involving social, economic, and environmental outcomes in urban and rural areas.
- Examine how to transform food systems to achieve critical progress on the SDGs and to contribute to a better future, including reducing inequalities and promoting well-being along economic, environmental, and social dimensions of sustainability.
- Examine the future of alternative proteins, precision fermentation, 3D printing of meat, fish, and plant-based proteins to support sustainable, nutritious, and equitable food systems including consumer acceptance.
- Explore the potential impacts of targeted technological innovations on urban and rural agriculture, agribusiness, food supply chain, animal welfare, climate change, energy, water, land use, biodiversity, health, and food loss and waste, as well as societal and cultural barriers they may encounter.

POSSIBLE ACTIONABLE STEPS FOR FOOD SYSTEMS

The committee proposes the following possible actionable steps to operationalize sustainable development in the area of sustainable and equitable food systems:

- Urban leaders and planners could be convened to discuss challenges, opportunities, and innovative strategies for sustainable and equitable urban food systems in the United States in the context of the COVID-19 pandemic, geopolitical conflict, and climate change. An example is the Milan Urban Food Policy Pact (2022), an international agreement on urban food policies signed by more than 200 cities from across the globe.

- Mayors and city officials could enhance cross-sector collaborations and engage communities to transform urban food systems rather than impose solutions from above or continue working in silos.
- Urban leaders could accelerate initiatives toward sustainable and equitable food systems with an appropriate sense of urgency, given that the urban population is projected to increase rapidly in the coming decades.
- Federal and local governments could work together to transform food systems in the United States and to promote supply chains that are resilient and sustainable in the short and long terms.
- Academic leaders and the private sector could support training and workforce development at universities, community colleges, and tribal colleges to promote sustainable and equitable food systems, taking into account environmental, economic, and social considerations.

5

Urbanization

Although Sustainable Development Goal (SDG) 11 most directly targets urban areas, cities will not realize the goal's description of becoming "inclusive, safe, resilient and sustainable" without achievement of related SDGs. Globally, of a population of 7.8 billion in 2020, 4.4 billion people lived in urban areas; by 2050, of a total projected population of 9.7 billion, 6.7 billion people are projected to live in urban areas (Figure 5-1; UN, 2019a). Eighty-three percent of the U.S. population lives in urban areas (Center for Sustainable Systems, 2022).

The role of urban areas in sustainable development has been increasingly recognized over the past several decades. Workshop presenter Marc Weiss of Global Urban Development recalled that at the 2002 UN World Summit for Sustainable Development, urbanization was barely on the agenda. Now, urbanization is recognized as an important phenomenon to address to operationalize sustainability. Improving the connections between urban areas and their intermediary cities and rural areas is also a growing area of research (OECD, 2021b).

CHALLENGES

Although local-scale sustainability transformations (e.g., restoring a park or lake, or creating a set of bike paths) are important, they are easier to achieve than systemic change across multiple dimensions of SDGs at the city-regional scale. Yet, setting large-scale science-based targets can have social and economic consequences. For example, decreasing carbon emissions by x percent, or increasing tree cover by y percent, may be possible, but doing so without exacerbating inequity or worsening poverty and vulnerability is more challenging and difficult. Although the rapid development of cities and influx of new residents have significantly

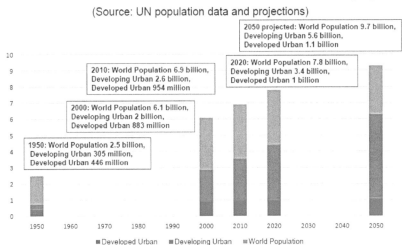

TOTAL WORLD POPULATION AND URBAN POPULATION, 1950-2050 (projected)
(Source: UN population data and projections)

FIGURE 5-1 Total world and urban population, 1950 to 2050.
SOURCE: Marc Weiss/Global Urban Development, Workshop Presentation, April 21, 2022, based on UN data (https://population.un.org/wup).

increased property prices and made cities unaffordable for many, efforts to reduce urban disparities will allow cities to become more sustainable and inclusive places to live and work. SDG 10 emphasizes that reducing inequalities and ensuring that no one is left behind are integral to achieving the SDGs (UN, 2022a).

CASE STUDIES AND SYNERGIES

Workshop presenters highlighted opportunities for synergies among SDGs related to urbanization. For example, restoring wetlands and urban forests can bolster food security, provide flood and drought relief, buffer urban heat island effects, and reduce air pollution, as well as provide city dwellers mental and physical relief from stress. Transitioning to low-carbon (e.g., bike-friendly or bus-based) transport systems can not only reduce carbon emissions, but also decrease obesity levels, improve local economies, and reduce air pollution. Globally, Copenhagen, Denmark (Box 5-1) and Porto Alegre, Brazil (Box 5-2) illustrate how these benefits can engage citizens to make sustainability fun and aspirational, not just scary and requiring sacrifice. It takes less time to commute by bike than to drive in Copenhagen, for example. Stormwater management areas that are part of the city's Cloudburst Management Plan are built as parks. Rather than "giving up privileges," it was suggested reframing the discourse when possible to "getting benefits" (Leonerdsen, 2022).

BOX 5-1
Global North: Copenhagen, Denmark

With an international reputation as a livable, relatively wealthy city, Copenhagen brings to mind images of people bicycling to work and of green spaces and parks. Lykke Leonardsen, program director of Resilient and Sustainable City Solution, shared that this environment was not always the case. She described the role of ambitious targets and co-benefits in engaging the population to create a more sustainable, resilient metropolitan area. As Leonardsen commented, "People see a city that was always rich. But this is not so. The inner harbor was heavily polluted 40 years go; people can now swim in it. In 1993, the city was bankrupt."

Focus on Livability

With this financial crisis came an opportunity to rebuild with livability linked to sustainability. The city's goal is to become carbon neutral by 2025. Much progress has been made, but Leonardsen acknowledged "the last mile is difficult." The key goals in the Copenhagen Climate Action Plan are to reduce energy consumption, reorient energy production to wind and other renewable sources, increase green mobility, and change how the city itself delivers results. Circular Copenhagen has set specific targets related to recycling, carbon dioxide reduction, and reuse of materials. Bicycling is part of everyday life. Significant for success, people do not bike necessarily with the environment as a motive but because it is the easiest and most convenient way to get around the city because of large investments in bicycle infrastructure.

Focus on Resiliency

Copenhagen expects that climate change will bring more extreme weather in the future. A Cloudburst Management Plan represents a comprehensive infrastructure effort with more than 300 projects to manage stormwater. The plan gives the city an opportunity to consider other issues; for example, water parks that can be used for water storage and recreation. The idea is to take advantage of climate adaptation with co-benefits: recreational value, biodiversity, social resilience, health, improved microclimate, accessibility and safety, and economic growth.

Five quality parameters that define urban nature in Copenhagen can be applied in other urban areas: biodiversity, climate adaption, functionality, spatial quality, and maintenance. Although an urban park or garden will differ from a forest in a more rural area, Leonardsen stressed the social and environmental benefits of green space. Rather than divorce urban dwellers from nature, nature should be integral to the development of a city. "It's just not fixing a problem, it's about creating a better city for citizens," she observed. "It's part of how to get citizens active and engaged." Rather than discuss hydrological issues, for example, residents can envision how the space could appear when it is not raining.

References

C40 Cities. Copenhagen. https://www.c40.org/cities/copenhagen.
Circular Copenhagen: Resource and Waste Management Plan. https://circularcph.cphsolutionslab.dk/cc/home.
Leonardsen, L. 2022. Workshop Presentation, April 21, 2022.

BOX 5-2
Global South: Porto Alegre, Brazil

Generating sustainable prosperity and quality of life for urban residents requires a new development paradigm, stated Marc Weiss, chairman and CEO of
Global Urban Development (GUD), but "fear and inertia get in the way." A solution:
"We need to replace economic fear with economic hope." GUD was founded in
2001 with a 30-year strategy to enable people to thrive in peace with each other
and with nature through sustainable innovation and inclusive prosperity.

"Four Greens"
Sustainable innovation economic development strategies are built on a paradigm that innovation, efficiency, and conservation in the use and reuse of all natural
and human resources is the best way to increase jobs, incomes, productivity, and
competitiveness (Nixon and Weiss, 2010; Weiss and Nixon, 2011). These economic
strategies are also the most cost-effective method of promoting renewable energy
and clean technologies, protecting the environment, and preventing harmful impacts of climate change. "People, places, and organizations get richer by becoming
greener," Weiss said, through:

- **Green savings:** Cutting costs of businesses, families, communities, and
  governments by efficiently using renewable resources and by reducing and
  reusing waste
- **Green opportunities:** Growing jobs and incomes through business development and expanding markets for resources efficiency, sustainability, and
  clean technologies
- **Green talent:** Investing in fundamental assets such as education, research,
  technological innovation, and modern entrepreneurial and workforce skills
- **Green places:** Establishing sustainable transportation and infrastructure,
  and protecting and enhancing the natural and built environment, to create
  more attractive, livable, healthy, productive, and resource-efficient areas
  and communities

Porto Alegre Sustainable Innovation Zone (ZISPOA)
GUD has been involved in a World Bank-funded strategy for the southern
Brazilian state of Rio Grande do Sul. To keep people actively engaged at the local
level and over the long term, the strategy employs Sustainable Innovation Zones
that combine six elements: (1) innovation and technology, (2) entrepreneurship
and startups, (3) sustainability and resource efficiency, (4) creativity and collaboration, (5) participatory community management, and (6) business-friendly
environment (Weiss, 2019).
ZISPOA was created within Porto Alegre, a city of 1.5 million. The aim is for
the city to become the most solar-powered, energy-efficient, bike-friendly, circular
economy, and digitally-connected community in Latin America by 2030. Community members have joined together to plan and take part in courses, marches,
festivals, community gardens, composting centers, electric car sharing, electric car
charging stations, solar posts and rooftops, sustainable parklets, bike sharing/bike
lanes/bike parking, and much more. Faculty and students in different disciplines

BOX 5-2 Continued

have assumed a leading role in these efforts. Groups that previously did not interact with each other now collaborate through ZISPOA.

A bottom-up approach, with neither state nor city officials in charge, involves civil society, academia, business, and government. Elements for change include taking action and producing results to show what sustainable improvements will look like; participatory inclusiveness; and independent nonpartisanship to survive electoral changes in political leadership. Professors, in addition to being experts, use their ZISProf platform to enable students and universities to become engaged in transformation. An interdisciplinary graduate Sustainable Innovation Professional university program is now offered. The ZISPOA concept has spread to Santo Angelo within Brazil, as well as to communities in Panama City, Panama; Western Sydney, Australia; and Poznan, Poland. GUD is also working in Wheaton/Montgomery County, Maryland, to develop a Sustainable Innovation Zone.

References

Global Urban Development websites, www.globalurban.net; www.globalurban.org.

Nixon, J. H., and M. A. Weiss. 2010. *Sustainable Economic Development Strategies.* Washington, DC: Global Urban Development. https://www.globalurban.org/Sustainable_Economic_Development_Strategies.pdf.

Weiss, M. A. 2022. Workshop Presentation, April 21, 2022.

Weiss, M. A. 2019. *The Porto Alegre Sustainable Innovation Zone (ZISPOA).* https://www.globalurban.org/ZISPOA_description_and_bibliography.pdf.

Weiss, M. A., and J. H. Nixon. 2011. *The Global Future of Green Capitalism.* Washington, DC: Global Urban Development. https://www.globalurban.org/Green_Capitalism.pdf.

ZISPOA website, https://www.zispoa.info.

The importance of collaboration and knowledge-sharing surfaces in almost any discussion about sustainable urbanization. A regional partnership spearheaded by the University of Texas at Arlington (Tare, 2022), public-private partnerships in New York City (Koval, 2022), and development of Voluntary Local Reviews (Saiz, 2022) served as useful examples for the committee. Efforts such as C40, Local2030 Hubs, and SDG Leadership Cities have created and strengthened communities of practice and knowledge-sharing. Yet, as several presenters warned, strategies must be participatory at all stages, not just in form but truly co-developed. If too closely tied with the agenda of a mayor or other leader at the helm, strategies could fall apart with changes in leadership.

It would be useful to examine issues of the expanding footprint of cities and the challenges of providing essential services to their residents in the Global North as well as the Global South. An example includes the Cities Development Initiative piloted by the U.S. Agency for International Development (USAID) in the Philippines as a strategic investment in secondary cities to foster larger regional development beyond the major urban cities (USAID, 2018).

KEY RESEARCH PRIORITIES FOR URBANIZATION

The committee proposes the following key priorities for research to operationalize sustainable development in the area of urbanization:

- Examine how to achieve systemic transformation across multiple dimensions of the SDGs at the city-regional scale while addressing transnational and rural-urban linkages and externalities, including shifting burdens (social, economic, and environmental) beyond the regional borders.
- Build a multiscale narrative of urban change that links local, national, regional, and global activities in the context of COVID-19, climate change, and global conflicts.
- Improve data collection and reporting at the local level, including of disaggregated and city-level data, as well as create open data hubs and portals to capture information from local agencies and community-generated data sets.
- Improve information on cities, including by filling data gaps, especially in the Global South, because developing cities will contribute the majority of the future urban transition in coming decades.
- Improve the understanding of the types of data needed from cities to monitor SDG transitions along environmental, social, *and* economic considerations. The first step would be to identify the types of data that are missing. The second step would be to determine how to collect these data across multiple locations.
- Explore how to conduct research on systemic equity and power that aligns with research on environmental and economic transitions, because this research is often conducted in silos.

POSSIBLE ACTIONABLE STEPS FOR URBANIZATION

The committee identified the following possible actionable steps to operationalize sustainable development in the area of urbanization:

- Urban leaders and planners could convene diverse, inclusive groups in workshops to focus on the key research priorities addressed above in the context of the COVID-19 pandemic, climate change, and global conflicts.
- Research institutions could create opportunities for workshop reports and journal special editions that focus sharply on identifying critical knowledge gaps relating to big data and research on cities and on producing new knowledge of special relevance to direct action, such as providing guidance to funders about areas for future work.
- International organizations could establish and maintain databases for international research on urbanization, such as the Urban Policy Platform (2022), that focus on urban-rural linkages and the Organisation for Economic Co-operation and Development work (2021b) that strengthens intermediary cities to achieve the SDGs.

6

Decarbonization

The largest contributor to greenhouse gas emissions through human activity is carbon dioxide (IPCC, 2022). Meeting the Paris Climate Agreement targets requires action throughout society, including the segments discussed throughout this report—food systems, transportation, energy usage, and more. Profound and exponential changes in human lifestyles, social institutions, governance, infrastructure, and technology are needed to meet the goals of the Paris Climate Agreement. Otto et al. (2020) propose key interventions to reduce greenhouse gases in this decade: "removing fossil-fuel subsidies and incentivizing decentralized energy generation, building carbon-neutral cities, divesting from assets linked to fossil fuels, revealing the moral implications of fossil fuels, strengthening climate education and engagement, and disclosing greenhouse gas emissions information" (Otto et al., 2020). In addition, the Intergovernmental Panel on Climate Change's Sixth Assessment (AR6) has identified the need to permanently sequester about 10 percent of current carbon emissions by 2050 to stay within the Paris temperature limits: that is, the sequestration of 6–10 gigatons/year by 2030 or sooner.

Carbon dioxide removal (CDR) technologies capture carbon either directly from the air or at a fossil-fuel source, then reuse or sequester it depending on the method. Both engineered and nature-based methods are at different stages of research, development, and deployment: direct air capture; mineralization; soils; forests; hybrid, such as bioenergy with carbon capture (BECCS); and ocean sequestration (Table 6-1).

TABLE 6-1 Pathways for Carbon Removal

Pathway	Summary
Direct air capture	Engineered CDR; early commercial deployment at ~4K tons/year with megaton scale plants planned
Mineralization	Engineered CDR; planned commercial deployment
Soils	Natural carbon sink
Forests	Natural carbon sink
Bioenergy with carbon capture	Hybrid CDR; not always removal/negative emissions
Oceans	Broad tech and non-tech opportunities; very early stage

SOURCE: Erin Burns, Workshop Presentation, May 16, 2022.

CHALLENGES

Decarbonization of energy systems is central to global decarbonization and achievement of all SDGs (IPCC, 2022; Nakicenovic, 2022; Nakicenovic and Lund, 2021). A fundamental energy-systems transformation would help to address health, climate, and other challenges facing humanity, and would especially benefit individuals without access to affordable and clean energy services (GCSA, 2021; IPCC, 2022; TWI2050, 2020).

The actions needed have been assessed multiple times over the past decades (GCSA, 2021; GEA, 2012; Häfele et al., 1981; IPCC, 2018; SAPEA, 2021). The first priority is to invest in decarbonization and efficiency because worldwide investments in renewables previously peaked in 2017 and have now reached new record highs because new renewables such as wind and photovoltaics can be less expensive than fossil alternatives (REN21, 2022). In fact, the investment costs of photovoltaic cells have declined by three orders of magnitude and are now lower than $1/Watt peak (GEA, 2012; SAPEA, 2021). A pervasive transformation toward zero-carbon electricity and electrification of end uses is central to decarbonization and net-zero emissions (IPCC, 2018). This effort should be complemented by low- and zero-carbon fuels such as hydrogen and CDR, along with sustainable biomass (BECCS) to achieve net-negative emissions. Major challenges to transforming energy systems include mobility that can be electrified through electric and plug-in vehicles, heating and cooling through heat pumps, and especially freight transport, aviation, and shipping. Blue hydrogen with CDR natural gas, and then green hydrogen and decarbonized synthetic fuels, offer a viable solution (GCSA, 2021; Nakicenovic, 2022).

Even the strongest advocates warn that CDR, no matter how robust and fully deployed, can never replace aggressive carbon reduction strategies and cannot be perceived as an alternative to mitigation (Burns, 2022). Moreover, CDR technologies are in the early stages of development, their unintended consequences are not known, and they require massive scale-up and financial investment to meet the AR6 goal.

Uncertainty related to performance, longevity, safety, and trust must be addressed. Developing laws and policies that account for timescale for long-term sequestration—up to millennia—is a similarly daunting challenge. Issues related to land-use; land ownership; and monitoring, reporting, and verification (MRV) must be resolved. The potential implications for biodiversity and society—especially given the uncertainties and the long timescale—remain unclear and the necessary data to inform decision-making are lacking. Sequestration does not create a "product," but rather a public good that requires public investment (Burns, 2022).

CASE STUDIES AND SYNERGIES

Decarbonization is both a technological challenge and a strategy with economic, environmental, and social consequences that are both known and unknown. Investment in research, development, demonstration, and deployment across the pathways summarized above (Table 6-1) has increased. A notable example is the $1 billion investment by the U.S. government to develop four regional Direct Air Capture hubs (DOE, 2022).

Researchers in academia, the nonprofit sector, and government are addressing the myriad of technical and nontechnical issues. For example, the committee discussed decarbonization options with representatives from Carbon180 (see Box 6-1), Carbon Clean (Bumb, 2022), and the International Biochar Initiative. As noted in Chapter 2, biochar offers potential benefits for soil and agriculture that warrant further investigation (Draper, 2022). Even though large-scale public investment is required to achieve decarbonization goals, the private sector is offering solutions, such as modular carbon capture and utilization technologies (Bumb, 2022), that open up new avenues of financing for CDR.

Certification can help ensure safety, performance, and trust. Yet, current certification and standards are inconsistent, incomplete, and lack rigor. A project coordinated through Arizona State University is examining existing schemes (Box 6-2). Regarding safety, the need for enormous volumes of storage will affect everyone, now and in future generations (Arcusa, 2022). Regarding performance, removal activities must function as promised. Regarding trust, carbon sequestration moves odorless, colorless gas that may have no discernable impact for years or decades. Moreover, certification must be conducted within a recognized and trusted framework. Potential consequences from improper certification include wasting time and resources, enabling scams/fraud, harming communities and the environment, and failing to address climate change.

Although CDR is emerging as an important climate agenda item, other decarbonization options will play a substantial role in reducing net emissions, including zero-carbon energy sources such as renewables and nuclear energy. Needed are efficiency improvements across the whole energy system, especially in end use, as well as new climate-friendly lifestyles and behaviors. As described in Box 5-1, Copenhagen aims to become carbon neutral by 2025 and to rebuild with livability linked to sustainability, by reducing energy consumption, reorienting energy production to wind and other renewable sources, and increasing green mobility. AR6 states that measures that promote walkable urban areas,

BOX 6-1
Connecting Policy, Action, and Science: Carbon180

Carbon180 was founded in 2015—the "180" refers to the need to "build a future that runs 180° from the past," explained Erin Burns, executive director. She stressed that, although the organization focuses on carbon removal at a gigaton scale, aggressive emissions reduction is also needed and must not be lessened as removal technologies improve. Carbon180 works with policy makers, entrepreneurs, and peer organizations across the United States to design policies that will bring necessary carbon solutions to the gigaton scale.

Scaling Challenges

Burns stated that each CDR pathway faces challenges related to costs, permitting, durability, and tradeoffs. No matter the pathway, two challenges have policy implications. First, carbon removal is not a product to sell, but fundamentally a public good. This characteristic shapes the types of policies needed to achieve scale. For example, different solutions are needed to reduce costs of solar energy in the short and long terms. Second, achieving the gigaton scale within decades requires coordinated research and deployment across carbon removal and reduction solutions, and different approaches may involve different capital costs.

Policy Solutions

Burns suggested policies to facilitate scale-up in six broad areas:

- Research, development, demonstration, and deployment (RDD&D). The reauthorization of the Farm Bill is a near-term opportunity to support the land-based pathways, and investment in direct air capture has accelerated. Federal coordination across agencies.
- International coordination through R&D, deployment, and incentives.
- Place-based strategies that optimize local resources and benefits for when and how to deploy CDR.
- Offering of incentives that prioritize key attributes, such as durability and looking forward, rather than prescription of a certain technology.
- Early purchasing and procurement. Federal government procurement as well as early purchases by private companies, such as Stripe and Shopify, have helped to catalyze the market.

References

Burns, E. 2022. Workshop Presentation, May 16, 2022.
Carbon180 website, https://carbon180.org.

combined with electrification and renewable energy, can improve health through cleaner air and enhanced mobility (IPCC, 2022). Design and management of urban areas play important roles in achieving decarbonization goals.

This grand transformation toward full decarbonization of energy systems and end use is not only about technology and economics. It is also about people, societies, and values and behaviors. Technology is an integral part of society and a

---

BOX 6-2
Certification of Carbon Removal and Sequestration

The role of certification is to provide direct (buyers) and indirect (public) assurance that a product, service, or person meets certain claims. Arizona State University (ASU), Thunderbird School of Global Management, and Conservation International have been involved in a multistakeholder collaborative project to develop a framework for carbon sequestration certification, reported Stephanie Arcusa, a postdoctoral fellow at ASU. The goal is to develop guidance for the development of evidence-based standards that lead to carbon sequestration today without compromising the needs of the future.

Why Certify?

Arcusa reported on a scoping analysis of the ecosystem of certification and standards for carbon sequestration today, which revealed more than 30 standards organizations and more than 125 standards for 23 types of activities to sequester carbon. Advances have been made, but none of the standards certify the same thing and they are not comparable. The certification ecosystem needs coherence across standards to ensure the credibility of the carbon removal industry.

Critical Challenges in Carbon Sequestration Certification

Unaddressed and urgent challenges include defining durable storage from a scientific basis and then devising business practice requirements accordingly; developing measurement techniques to be used in accounting protocols with acceptable levels of uncertainty; identifying what constitutes a suitable reservoir and the consequences of adding carbon to reservoirs; and establishing independent verification. She noted the most important need is to convincingly certify durable storage.

References

Arcusa, S. 2022. Workshop Presentation, May 16, 2022.
Arcusa, S., and S. Sprenkle-Hyppolite. 2022. Snapshot of the carbon dioxide removal certification and standards ecosystem (2021–2022). *Climate Policy.* https://doi.org/10.1080/14693062.2022.2094308.
Global Carbon Removal Partnership website, https://www.carbonremovalpartnership.net.

---

collective expression of sundry individual choices (Nakicenovic and Lund, 2021; Nakicenovic, 2022).

KEY RESEARCH PRIORITIES FOR DECARBONIZATION

The committee proposes the following key priorities for research to operationalize sustainable development to contribute to decarbonization:

- Examine fundamental science for ocean- and nature-based CDR, including chemical pathways, microbiome variability and durability of soil sequestration, forest and ocean-based proposals, and suitable reservoirs for underground or deep sea storage.[1]

- Conduct standards setting for monitoring, reporting, and verification techniques for various pathways.
- Explore acceptable levels of uncertainty in certification in both technical and social dimensions including intergenerational justice.
- Improve the understanding of possible impacts on biodiversity, land, or ocean use for food or other unintended consequences such as tipping points for carbon sinks becoming sources.
- Examine technologies that enable large-scale deployment of carbon capture, utilization, and storage, with an emphasis on durability and ways to scale-up. Examples include the Sleipner T carbon dioxide treatment platform and the carbon capture plant in Iceland (Panko, 2021).

POSSIBLE ACTIONABLE STEPS FOR DECARBONIZATION

The committee identifies the following possible actionable steps to operationalize sustainable development to contribute to decarbonization:

- The U.S. government and other national governments could identify strategies for CDR that are place-based, community embraced, and environmentally and intergenerationally just.
- The U.S. government could build on the current U.S. $1 billion allocated for the four regional Direct Air Capture hubs (DOE, 2022) to establish other CDR demonstration projects, such as biochar in concrete, asphalt, and soil, as well as global satellite forest monitoring.
- The U.S. government and other national governments could set a flue point capture target of $50/tonne for hard to abate industries, including establishing an international prize competition. An example includes the $100 million Prize for Carbon Removal initiative (XPrize Foundation, 2022).
- The U.S. government and other national governments could ramp up research, development, demonstration, and deployment (RDD&D) for all forms of CDR, as companies are increasing their investments for decarbonization efforts (Judge, 2022).
- The U.S. government and other national governments could play a leadership role in international collaboration and co-funding of research, provide international incentives for ethical deployment and scale-up, and propose an international framework for standards and MRV to deter

---

[1] The National Academies has released a number of in-depth studies on many of these topics, including NASEM. 2022. A Research Strategy for Ocean-based Carbon Dioxide Removal and Sequestration. Washington, DC: The National Academies Press. https://doi.org/10.17226/26278; and NASEM. 2019. Negative Emissions Technologies and Reliable Sequestration: A Research Agenda. Washington, DC: The National Academies Press. https://doi.org/10.17226/25259.

national and corporate "greenwashing." Although groups such as Frontier[2] and First Movers[3] are using procurement strategies to accelerate carbon removal, standards are needed. Relevant examples include Climate Bonds Certification (2022) and Green Bond Principles (2022). Bilateral city exchanges, city networks, and academic institutions could encourage collaborative research opportunities in emerging countries.

- The U.S. government could enhance federal coordination between agencies, such as the U.S. Department of Agriculture, Department of Energy, Department of Interior, Environmental Protection Agency, General Services Administration, and Department of Defense relating to incentives, MRV, siting, and accounting.

- The U.S. government could expand attribute-focused rather than prescriptive tax incentives, such as focusing on sequestration durability instead of pathway-specific technology.

- Federal and state governments and international coalitions, working with multi-sector partnerships such as the Global Carbon Removal Partnership,[4] could use procurement to catalyze and set standards for private-sector investments, promote incentives for sequestration and not just capture, and utilize lessons learned from partnerships (see case studies) to engage in international dialog on ethics and environmental justice in CDR, as well as framework and standards for MRV.

- Governments, the private sector, and nongovernmental organizations could work together to promote decarbonization in agriculture, industry, and energy production, including by building carbon-neutral cities, strengthening climate education and engagement, and encouraging low-carbon lifestyles for mobility, housing, and consumption.

---

[2] See https://frontierclimate.com.

[3] See https://www.protocol.com/first-movers-coalition-climate-davos.

[4] See https://carbonremovalpartnership.net.

7

Science, Technology, and Innovation for the Sustainable Development Goals

Seven years after agreeing on the Sustainable Development Goals (SDGs), the world is not on track to achieve them, and overlapping crises (including COVID-19, geopolitical conflict, and climate change) threaten the progress made (Mueller, 2022). Applying science, technology, and innovation (STI) across the SDGs may represent a way to recalibrate and move ahead. Advances in digital technologies, renewable energy, and other technologies continue to offer opportunities to achieve the SDGs. In recognition of this potential, the United Nations (UN) has convened a Multi-Stakeholder Forum on Science, Technology, and Innovation for the Sustainable Development Goals (STI Forum) for the past seven years. The committee held a side event workshop at the most recent gathering in May 2022.[1]

CHALLENGES

Several challenges to applying STI have surfaced and, in some cases, have been exacerbated by the COVID-19 pandemic and political and social unrest. One such challenge is the digital divide, in which access to technology is uneven and inequitable across and within countries (NASEM, 2022b). As the interconnected world relies more heavily on digital technologies, countries and people without this access to them may fall further behind (Tilmes, 2022).

In addition, the scope and reach of social media, artificial intelligence algorithms, and government and private control of media have implications for personal privacy and freedoms. Full realization of the benefits of technology and mitigation of its detriments require appropriate governance, infrastructure,

---

[1] For the seventh annual STI Forum, see https://sdgs.un.org/tfm/STIForum2022.

resources, and capabilities, as well as the capacity of individuals, communities, and companies to absorb and apply them. STI are major pillars for accelerating progress toward the SDGs. Research, development, deployment, and widespread diffusion of environmentally sound technologies are essential to advancing sustainable development (UN, 2022b).

CASE STUDIES AND SYNERGIES

The African continent provides examples of the potential for digital technologies (Adam, 2022). Three promising areas of development relate to the SDGs:

- Digital strategies can allow countries to leverage leapfrogging opportunities and achieve a faster rate of development than they would otherwise. Several start-ups are using blockchain technology to facilitate transportation, traceability, and input linkages. For example, the Ghanian start-up company Bitsika provides a secure, inexpensive way to send and receive money between African countries. Mobile phones enable business transactions, communication, payments, and lending. Much business in Africa is conducted in the informal sector, especially among women, but 80 percent of the population has a mobile phone, which can serve as a building block for economic inclusion.
- Pharmaceutical manufacturing and production within Africa ramped up quickly to source personal protective equipment and devices needed during the COVID-19 pandemic. A digital platform allowed countries to post their requirements and companies to bid to fulfill those requirements. This platform boosted local production and financing, was upscaled quickly, and could be adapted to other sectors.
- Carbon-neutral energy production has significant potential, with 600 million Africans without access to electricity and with many countries holding critical resources. For example, 70 percent of the world's supply of cobalt, which is critical for renewable energy and electric vehicles, is sourced from the Democratic Republic of Congo. According to Adam, studies show that investing in local (African) manufacturing can lead to battery precursors that are 30 percent less expensive than those manufactured in the United States and China. The core goal is to increase the opportunity for investment as part of a revolution of production.

An important focus on STI for the UN 2030 Agenda is the UN Technology Facilitation Mechanism (TFM). The TFM has four components: (1) an annual STI Forum as described above; (2) the 10-Member Group (Group of High-level Representatives of Scientific Community, Private Sector and Civil Society) appointed by the UN Secretary-General for two-year terms; (3) the IATT (UN Interagency Task Team on Science, Technology and Innovation for the SDGs) with one representative from each of more than 40 international agencies; and (4) 2030 Connect (an online platform for existing STI initiatives, mechanisms, and programs) (UN, 2022d). STI for SDGs roadmaps, led by the IATT and other groups, offer a promising approach to planning for how STI can accelerate a country's effort toward the SDGs (Box 7-1).

STI partnerships across sectors and disciplines offer hope for resurgent multilateralism and innovative approaches to advance the SDGs (Truman Center, 2022). This includes strategic collaboration among the science community, such as the National Academies' Nobel Prize Summit (NASEM, 2021a), and through initiatives such as the International Science Council's "Sustainability Science Missions" (ISC, 2021) and their synthesis of research gaps; the six transformations in The World in 2050 Report (Box 7-2); and the New European Bauhaus initiative, an interdisciplinary initiative that connects the European Green Deal to our living spaces and experiences (EU, 2022). The European Union has also promoted "innovation cohesion," a new approach to support industrial research and development (R&D) to better connect research and innovation stakeholders in Europe (Zubașcu, 2021). Further, emerging scientific fields (White House, 2022) offer the benefits of multi-scalar analysis using interoperable data sources.

Partnerships are emerging between city networks and the STI community to serve as intermediary knowledge brokers and catalytic technical advisors to support innovation from the local to global scale. Examples includes the climate action network of megacities known as C40, and other networks in which local governments work collectively to advocate and implement shared initiatives across borders.

A significant amount of progress has been made on sustainable design and operation of technologies for buildings, transportation vehicles, transportation infrastructure, energy production, and metal working through science, technology, engineering, and innovation. As described in Chapter 6, Carbon Clean is delivering cost-effective carbon capture technology to achieve net zero through industrial decarbonization. For example, its CycloneCC™ technology enables scalable cost-effective carbon capture for the industrial sector by reducing equipment size and CapEx and OpEX up to 50 percent (Bumb, 2022). In more than 44 facilities and references across the globe, Carbon Clean has captured more than 1.5 million tonnes of carbon dioxide since 2009 through strong global partnerships.

The number of publications highlights the critical role of engineering in empowering sustainable development and achieving many of the 17 SDGs, such as addressing poverty, supplying clean water and energy, responding to natural disasters, and constructing resilient infrastructure (UNESCO, 2021). The role of engineering is essential for developing livable, sustainable, and resilient cities that incorporate energy efficiency into buildings, efficient transportation systems, and effective infrastructure and water resource management.

BOX 7-1
STI for SDGs Roadmaps

One promising approach to accelerating the process of developing new or adapting existing solutions to meet the Sustainable Development Goals (SDGs) and their targets is national development of "roadmaps." The concept of a road-map is familiar to many people who work in the technology sector. In this case, what are known as STI4SDG Roadmaps help a country to build ownership and coherence across its development plans, SDG programs, and science, technology, and innovation (STI) initiatives. IATT began this pilot project with Ethiopia, Ghana, India, Kenya, and Serbia and continued to scale with the addition of Ukraine. Working together with the 10-Member Group, the Government of Japan, the World Bank, EU's Joint Research Centre, and the Organisation for Economic Co-operation and Development, IATT released a guidebook for the preparation of STI for SDGs roadmaps in 2021 (UN, 2021a).

Developing the Roadmap
The Roadmap has six steps: (1) define objectives and scope; (2) assess the current situation; (3) develop vision, goals, and targets; (4) assess alternative pathways; (5) develop detailed STI for SDGs roadmap; and (6) execute, monitor, evaluate, and update the plan (UN, 2021a). Key inputs include stakeholder consultations, technical and managerial expertise, and data and the evidence base.

Lessons from the Roadmap Process
- Ensure active participation across government and stakeholder groups to develop a coherent vision, goals, and targets (UN, 2021b);
- Employ up-to-date data and expertise to assess STI options; and
- Earmark budgets to implement the initiative.

Two of the pilot countries for STI for SDGs roadmaps, Ethiopia and Ukraine, are involved in conflicts that have disrupted their plans for sustainable development, illustrating that wars may be the greatest threat to achieving the SDGs (APS, 2022) as described in Chapter 7.

References

APS (American Physical Society). 2022. July 2022 Newsletter. https://engage.aps.org/fps/resources/newsletters/july-2022.

Colglazier, E. W. 2018. The Sustainable Development Goals: Roadmaps to progress. *Science Diplomacy* 7(1). https://www.sciencediplomacy.org/editorial/2018/sdg-roadmaps.

Tilmes, K. 2022. Workshop Presentation, May 4, 2022.

UN. 2021a. Guidebook for the Preparation of Science, Technology, and Innovation (STI) for SDGs Roadmaps. https://sdgs.un.org/sites/default/files/2021-06/GUIDEBOOK_COM-PLETE_V03.pdf.

UN. 2021b. Progress Report of the Global Pilot Programme on STI for SDGs Roadmaps. https://sdgs.un.org/sites/default/files/2021-04/Progress%20Report%20of%20Global%20Pilot%20Programme%20of%20STI%20Roadmaps_2021_1.pdf.

BOX 7-2
STI for Sustainability Transformation

The World in 2050 (TWI2050) by the International Institute for Applied Systems Analysis (IIASA), the Sustainable Development Solutions Network (SDSN), and the Stockholm Resilience Centre (SRC) has suggested a blueprint for six transformations needed to achieve the Sustainable Development Goals (SDGs) and long-term sustainability to 2050 and beyond. The six areas are (1) human capacity and demography; (2) consumption and production; (3) decarbonization and energy; (4) food, biosphere, and water; (5) smart cities; and (6) digital revolution (TWI2050, 2018; Figure 7-2-1).

Each transformation describes a major change needed to transform resource use, institutions, technologies, and social relations to achieve key SDG outcomes, and employs a people-centered perspective to promoting poverty reduction, fair distribution, and inclusiveness necessary for human prosperity. The report emphasizes the need for urgent, ambitious, and vigorous actions, a holistic perspective, and effective and inclusive governance for sustainability transformation (TWI2050, 2018). The TWI2050 initiative also developed two additional reports on the digital revolution and sustainable development (TWI2050, 2019) and innovations for sustainability with the crucial role of governance for SDG implementation (TWI2050, 2020).

FIGURE 7-2-1 Six Transformations.
SOURCE: TWI2050, 2018. Courtesy of Attribution-NonCommercial 4.0 International (CC BY-NC 4.0).

BOX 7-2 Continued

References

Tilmes, K. 2022. Workshop Presentation, May 4, 2022.

TWI2050 (The World in 2050). 2020. Innovations for Sustainability. Pathways to an Efficient and Post-Pandemic Future. Report prepared by The World in 2050 initiative. International Institute for Applied Systems Analysis (IIASA), Laxenburg, Austria. http://pure.iiasa.ac.at/id/eprint/16533.

TWI2050. 2019. The Digital Revolution and Sustainable Development: Opportunities and Challenges. Report prepared by the World in 2050 initiative. IIASA, Laxenburg, Austria. http://pure.iiasa.ac.at/id/eprint/15913.

TWI2050. 2018. Transformations to Achieve the Sustainable Development Goals. Report prepared by the World in 2050 initiative. International Institute for Applied Systems Analysis (IIASA), Laxenburg, Austria. http://pure.iiasa.ac.at/15347.

STI is a way to engage youth in development issues, and could provide an opportunity to enhance training and capabilities of a technology-savvy workforce in both formal and informal settings. In an ever-evolving science and technology landscape, this approach will be needed to enable transformations in a variety of fields: from food systems, to decarbonization, to data transparency and validation, among other topics addressed by the committee. Looking ahead, certain enabling emerging technologies such as quantum sensors may help overcome existing barriers of classic light-sensing technologies (Casacio et al., 2021). The nondestructive nature of this technology could enable long-term interrogation of environmental systems under dynamic conditions over an extended period of time. Other applications in biomedicine, energy, material science, engineering, environmental monitoring, and sustainability are possible and deserve additional investigation.

KEY RESEARCH PRIORITIES FOR STI FOR THE SDGS

The committee proposes the following key priorities for research to operationalize sustainable development in the area of STI cooperation:

- Examine the current status of achieving the SDGs in the United States and what resources and actions are needed to advance the SDGs in the context of the economic crisis, the COVID-19 pandemic, and geopolitical conflicts, building on the SDGs Report by UN Development Programme (2022c) and the Brookings and UN Foundation's report on The State of the Sustainable Development Goals in the United States (Pipa et al., 2022).
- Measure success of STI partnerships, such as the STI for SDGs roadmaps as described above. Simply counting the number of participants or partnerships does not equate to progress or stronger multilateralism.

- Explore what voice should cities, city networks, and their partners have in the multilateral system and how they can impact future international commitments.

POSSIBLE ACTIONABLE STEPS FOR STI FOR THE SDGS

Evidence demonstrates that partnerships across sectors and disciplines, including in STI, offer hope for resurgent multilateralism and innovative approaches to advance the SDGs. The committee identifies the following possible actionable steps to operationalize sustainable development in the areas of STI for the SDGs:

- Governments, international organizations, nongovernmental organizations, the private sector, and scientific communities could discuss a new vision for a sustainable and resilient future beyond the 2030 Agenda for Sustainable Development, because the world is not on track to achieve the original 2015 targets by 2030.
- Governments, the private sector, and nongovernmental organizations could look to all sectors (not just national governments) for partnerships that consider ways to have a greater voice in shaping our multilateral system and future shared commitments.
- Governments, the private sector, and nongovernmental organizations could become involved in the efforts at the UN to assist volunteer countries in developing their STI for SDGs roadmaps and to facilitate knowledge exchange and transfer at the local level.
- The UN could maintain the continuity of experience for its various scientific groups, such as the 10-Member Group and Global Sustainable Development Report (GSDR) group, which might play an important role in the post-2030 processes. Both groups recently appointed new members after two-year terms, which might impact continuity.
- Leaders and practitioners who have participated in the 10-Member Group, GSDR, and other UN SDGs efforts could share knowledge and experiences with current members or scientific groups to ensure the continuity of knowledge and engage with young people.
- The private sector could lead the advancement of STI in sustainable design and operation toward scalable achievement of the SDGs and influence the sustainability profile of multiple economic sectors.
- Companies could enhance public-private partnerships to support disaster recovery through STI.
- Philanthropic organizations could highlight and support examples of effective solutions to SDG challenges and sustainability challenges at the local, national, and global levels.

8

Science and Peace

The committee undertook its work in the context of war in Ukraine and increasing violence in conflict, criminal, and interpersonal settings around the world. In the first months of 2022, more than 100 million individuals were "forcibly displaced worldwide as a result of persecution, conflict, violence, or human rights violations" (UNCHR, 2022). Conflict undermines achievement of all Sustainable Development Goals (SDGs); conversely, climate change and other environmental changes implicit or explicit in the other SDGs affect conflict and social upheaval. Many crises have roots in economic and power inequalities, overexploitation of natural resources and people, and the struggle to control access to resources (e.g., fossil fuels, water, rare metals, land). The average prevalence of bribery is five times higher in low-income countries (37.6 percent) than in high-income countries (7.2 percent), which affects citizens interacting with essential public services such as health care, education, water, electricity, judiciary, and police (UNDOC, 2021). The war in Ukraine, as well as increasing conflicts and tensions in other world areas, highlight the need to reinvigorate the international institutional order that was established after World War II to ensure peace.

As the committee learned during a workshop session, SDG 16—Peace, Justice, and Strong Institutions—was not a given on the global sustainable development agenda (Lilja, 2022). Some of the countries most affected by violence and conflict demanded inclusion of this goal among the set of SDGs. A search for indicators and targets that made sense across states led to consideration of both "negative" (absence of violence) and "positive" (access to justice and inclusive decision-making) peace-related targets.

CHALLENGES

Peace and lack of violence have been sadly elusive in the 21st century, as reflected in the challenges to meet this goal and its targets. Challenges include ownership to implement and monitor the goal, with a disconnect between global and national levels. Many of the people who are charged with meeting the goal were not part of the process to develop it (Lilja, 2022). Another challenge is the multisectoral nature of the goal. Although all of the SDGs are necessarily cross-cutting, SDG 16 has a particularly diverse constituency that encompasses child rights organizations, peace-building organizations, police, military, democracy groups, judiciary, and many others. SDG 16 covers a lot of ground, including the reduction of all forms of violence, equal access to justice for all, increased accountability and transparency, and the protection of fundamental freedoms.

SDG 16 is one of the weakest in terms of data availability and monitoring (Lilja, 2022). The diversity of activities and stakeholders results in poor or out-dated data, as well as lack of geo-localized and subnational-level data (Basnyat, 2022). An SDG 16 hub on data collaboration was created, driven by a few experts, but is no longer active (Lilja, 2022). Data and information access that is closely connected with education access is important for facilitating independent thinking among citizens (Campbell, 2006). Many authoritarian regimes and dictatorships all over the world change school programs, restrict the freedom of press, control the media and the internet, and persecute opposition leaders. Attention tends to be focused on conflict between nations or regions, but SDG 16 also refers to more common yet still devastating forms of criminal and interpersonal violence (Locke, 2022). More knowledge is needed about individual and group trauma and its influence on individual and group behavior violence, today and for future generations (Locke, 2022).

CASE STUDIES AND SYNERGIES

Although SDG 16 involves a richness of targets and indicators, the committee's information-gathering workshop focused on its reducing violence aspects, especially on science and peace. The lack of data emerged from almost every workshop session, but seemed particularly salient during this session. To underscore this challenge, SDG 16 has 24 indicators; many were established despite the absence of globally agreed-upon methodology or data, although some improvements in methodology, even with limited data, have been noted (Basnyat, 2022). The availability of data at both the national and subnational levels can influence national policymaking. Having a global framework helps to prioritize collecting, unpacking, and analyzing data. The Stockholm International Peace Research Institute is hosting a consortium of data programs to use the best data on conflict to track violent deaths, in collaboration with several partners, and to demonstrate how to combine datasets from multiple sources (Lilja, 2022). The United Nations (UN) Development Programme is also working with countries to achieve the

BOX 8-1
Partnerships for Data to Achieve SDG 16

Partnerships are critical for the massive effort to collect data related to SDG 16. Because of the joint effort of national governments, civil society, and international actors, all indicators under SDG 16 now have internationally recognized methodologies, which was not the case when the framework was adopted in 2015. However, data are still not consistently collected at the global, regional, and national levels for many of the indicators.

To address this gap, partnerships across the UN system, such as through the *SDG 16 Survey Initiative*, have been critical to bring together national actors to promote data collection and reporting on indicators related to justice, human rights, and governance. For example, Tunisia's pilot of the SDG 16 Survey 2021 promoted collaboration between civil society and government at all levels to analyze the findings from the pilot, identify policy recommendations, and link to policymaking processes.

Although the SDG 16 indicators can help illustrate some aspects of peace, justice, and inclusive institutions, the broader field of governance statistics remains a relatively new domain of official statistics. The Praia Group on Governance Statistics, established in 2015, is working to further elaborate standards on governance statistics. Under the Praia Group, two task teams composed of a range of national and international experts and statisticians are currently working to pioneer global methodologies to measure "non-discrimination and equality" and "participation in public and political life." Future task teams will explore measurement of additional domains of governance including openness, access and quality of justice, responsiveness, absence of corruption, and trust.

Several strong practices to promote data collection and reporting across countries were highlighted during the workshop:

- In Guatemala, adoption of a multistakeholder approach to the development of the Voluntary National Review led to an increase in data availability, buy-in, and coordination.
- In the United Kingdom, the National Statistics Office created an open national reporting platform for the SDGs that is free to use and customize by other countries.
- In South Africa, the Centre for Human Rights of South Africa developed a training curriculum for university students on disability inclusion, including a module on Access to Justice, which is now being used in nine other countries.
- In Canada, Statistics Canada has invested in developing the methodology and producing data on gender identity in an effort to promote the inclusion of all.

References

Basnyat, A. 2022. Workshop Presentation, May 16, 2022.

UN Statistics Division. 2022. Praira Group on Governance Statistics. https://unstats.un.org/unsd/methodology/citygroups/praia.cshtml.

UNDP (United Nations Development Programme), UNODC (United Nations Office on Drugs and Crime), and OHCHR (Office of the United Nations High Commissioner for Human Rights). 2022. SDG16 Survey Initiative: Implementation Manual. https://www.undp.org/publications/sdg16-survey-initiative-implementation-manual.

SDGs (Box 8-1). The literature on "fragile and conflict afflicted states" is extensive (Watkins, 2018). Another collective effort to generate data relevant to SDG 16 targets and indicators is the World Justice Project (World Justice Project, 2022). This multisector collaboration around the SDG 16+ agenda provides a roadmap for peaceful, just, and inclusive societies (NYU, 2022).

In each geographic context, an ecosystem of actors is involved in peacebuilding. Each actor experiences different levels of credibility, leverage, and trust. Ideally, these levels should be reflected in who gets funding. When this does not happen—for example, because the local stakeholders may be unknown in multilateral headquarters or donor capitals—intermediary groups can provide the bridge, again highlighting the value of localization as discussed in Chapter 5. The global scientific community can play an important role in peacebuilding efforts. As a concrete example, the Polish Academy of Sciences has been helping Ukrainian scientists who are refugees or remained within their country (Slowinski, 2022). Other examples include global efforts of the World Academy of Sciences (UNESCO-TWAS), the InterAcademy Partnership, and the International Science Council to advocate for displaced scientists worldwide (IAP, 2022); National Academies efforts to support the resettlement of scientists and engineers from Afghanistan (NASEM, 2022c); and the Breakthrough Prize Foundation's and the U.S. National Academy of Sciences' efforts to support displaced Ukrainian scientists (NASEM, 2022a).

KEY RESEARCH PRIORITIES FOR SCIENCE AND PEACE

The committee proposes the following key priorities for research to operationalize sustainable development in the area of science and peace:

- Strengthen SDG data hubs, partnerships, and data for SDG monitoring and enforcement relating to science and peace and other relevant issues. The data are collected not for the sake of data but to determine how they can effect change at the local and national levels.
- Explore survey instruments on SDG 16 and the interlinkages between different variables while supporting countries to collect data on access to justice, corruption, discrimination, and trafficking. UN agencies are working on survey instruments that cover most of the SDG 16 indicators (UNDP, 2022a). Another good example of practice in this context is the Praia Group on Governance Statistics, which is led by statistical offices but invites participation by many different experts. The Praia Handbook was launched in 2020 (Basnyat, 2022).
- Examine how to deal with post-conflict trauma because global conflicts will influence future generations, including the war in Ukraine.
- Prevent and mitigate the effects of child soldiers and gender-based violence that occurs within conflicts (Stohl, 2018; UNICEF, 2021).

POSSIBLE ACTIONABLE STEPS FOR SCIENCE AND PEACE

Science, technology, and medical communities could contribute to world peace now more than ever. The committee proposes the following possible actionable steps to operationalize sustainable development in the areas of science and peace:

- Governments and nongovernmental organizations across the globe could be brought together, perhaps as part of the effort to negotiate the initiative that follows the SDGs beyond 2030, to design a new global social compact that emphasizes peace building and can promote a global system that builds on care, sharing, sufficiency, and respect for human and non-human living beings.
- Leaders and practitioners in governments, academia, nongovernmental organizations, and international organizations could create peer groups for implementing and monitoring the SDGs (e.g., cities learn best from other cities), dealing with crisis situations, and facilitating exchange among justice actors, peace builders, and inequality experts (Locke, 2022).
- The scientific community could promote positive examples for supporting Ukrainian scientists, and additional efforts and funds are needed to support science, engineering, and medical professionals in other nations including Afghanistan, Myanmar, South Sudan, Syria, Venezuela, Mexico, and Nicaragua (https://scienceinexile.org). Although the Institute of International Education's Scholar Rescue Fund supports fellowships for threatened and displaced scholars worldwide (IIE, 2022), additional efforts to protect people and the planet are needed.
- The U.S. government could be reoriented as the champion of SDGs through participation in a Voluntary National Review (see Chapter 3). It is essential to advocate for the importance of peace and justice (Locke, 2022).
- Funding agencies and philanthropic organizations could invest in organizations between state and society that can contribute to solutions that address the increasing number of wars, conflicts, and migration of displaced people. It is essential to support scientific communities in long-term sustainable development, including rebuilding efforts.
- Scientific societies and academies could conduct studies and dialogues that help to advance control of new and emerging weapon systems as has been the case with nuclear weapons.
- Nongovernmental organizations could support humanitarian efforts in conflict areas. A positive example is the World Central Kitchen, which delivers food and provides shelter to children and families in Ukraine (World Central Kitchen, 2022).

9

Financing to Achieve the
Sustainable Development Goals

More than $12 trillion in private investment is needed globally to address the challenges to meeting the Sustainable Development Goals (SDGs) by 2030 (Business and Sustainable Development Commission, 2017), in addition to the public investment suggested throughout this report. The World Bank estimates that low- and middle-income countries face investment needs of $1.5 trillion to $2.7 trillion per year (4.5–8.2 percent of their combined gross domestic product [GDP]) between 2015 and 2030 to meet infrastructure-related SDGs that depend on policy choices (Tippet, 2020; Vorisek and Yu, 2020). Accelerating global progress on the SDGs requires access to this capital and will necessitate new models and partners to finance projects and transitions, as well as greater openness to new customers and a willingness to redefine risk.

CHALLENGES

Financing can support realization of many of the opportunities identified throughout this report, including food system transformations, carbon dioxide removal, and sustainable urbanization. Yet, unlocking this capital is complex. Public companies are accountable to their shareholders. Start-ups are pioneering exciting innovations but with limited capital. The places that need the most investment are often those with the most limited access to capital. Many subnational governments lack creditworthiness or access to capital markets and/or may not be able to issue debt to finance needed and transformational projects. Municipal financing is constrained by limited access to capital markets for transformational projects while balancing emerging economic opportunities with environmental risks (UN, 2019b). The risk of inaction, including the impact on financial investments, must be taken into account. There is a need for innovation in microfinancing for small-scale SDG-related projects, and the U.S. Forest Service's Community Forest Program could serve as a model for rural and indigenous communities (USFS, 2022).

CASE STUDIES AND SYNERGIES

Despite these challenges, opportunities exist to realize tangible and intangible benefits from SDG-related investing. Place-based initiatives can be attractive investments for private capital, whether alone or through public-private-philanthropic (P3) partnerships. For example, Invest NYC SDG has attracted the private sector to sustainable investments in one of the most investor-focused cities on the planet (Box 9-1). The demand for positive ESG (environmental, social, and governance) investments creates demand for more blended finance options that are favorable to public, private, and other projects with significant and measurable social benefits. This presents opportunities for investing alone, through new P3s, or through other partnerships. Impact investing, development aid, and concessionary capital could jumpstart a project or a partnership that provides access to previously underserved markets. As one example, Hawai'i Green Growth was intentionally designed as an economic growth strategy (see Chapter 3).

Workshop presenters shared strategies to increase financing to achieve the SDGs. First, cross-sector collaboration is essential to unlock the combined capacity of the private sector, public sector, and community. All relevant stakeholders including local communities must be at the table to collaboratively set goals, commit to action, and agree on metrics to assess results. Second, blended finance offers an opportunity to increase the amount of overall resources while offsetting some risk (OECD, 2021a). The private sector understands that investments take time to put in place in other sectors; this mindset should apply to sustainability projects as well, and multisector partnerships that include government investment may help to remove regulatory impediments. Examples are numerous (Table 9-1). Across the six sectors in the table, the greatest level of innovation and capital mobilization has been observed in the renewable energy and the built environment sectors (NYU, 2021).

Through partnerships such as the United Nations (UN) Global Compact (and Global Compact USA), private-sector companies are setting ambitious targets for themselves and their peers (Gordon, 2022). This is made measurable and comparable through benchmarks developed by the World Benchmarking Alliance (Box 9-2) as one aspect of operationalizing the SDGs. The private sector is also working to establish integrated reporting standards for the ESG measures to which companies align their strategic planning, investments, and disclosures. A shared, transparent, and global set of standards for ESG reporting and evaluation is emerging from the Value Reporting Foundation,[1] a consolidation of ISSB and

---

[1] The Value Reporting Foundation merged with the International Financial Reporting Standards Foundation in August 2022. See https://www.ifrs.org/issued-standards/ir-framework.

BOX 9-1
Engaging with the Private Sector: Invest NYC SDG

New York City (NYC) has been a leader in aligning its sustainability plans with the Sustainable Development Goals (SDGs), beginning with the country's first Voluntary Local Review. The New York University Stern Center for Sustainable Business saw the opportunities to support implementation, reported Marianna Koval, and launched a place-based initiative in the city to develop projects to advance the United Nations (UN) SDGs and attract private investment.

Invest NYC SDG

Invest NYC SDG began with $760,000 in funding for two years. It is now funded for five years and has raised a total of $3.5 million with a small staff, bolstered by undergraduate and graduate student teams, executives, and fellows (including retired experts and others who have provided pro bono advice). A governing board of 14 people includes UN, business, government, and community representatives. Koval noted the center is within an academic institution, but it is not a research institution, an incubator, or an investor. "We recognized our value as a catalytic matchmaker, identifying opportunities and connecting."

Six focus areas have been mapped against the SDGs: sustainable mobility, built environment, waste, renewable energy, climate resilience, and food and health. The center developed criteria to identify high-impact projects that could attract private investment and metrics to measure success, inspired by the book *Drawdown*. Convenings brought sectors together. As Koval pointed out, the private sector cannot do it alone; the sustainable goals are a political document and the city controls the infrastructure. But the city will not be successful without the private sector. COVID-19 and the murder of George Floyd galvanized the community. Invest NYC SDG made economic equity core to all efforts, which include the following:

- **Riker's Island:** This 413-acre island in the East River currently has 10 jails housing 5,000 people. Considered a symbol of injustice, the facility is scheduled to close in 2027. Invest NYC SDG is helping to reimagine this large tract of city-owned land as a site of growth and regeneration through urban agriculture.
- **Equitable Commute Project:** During the pandemic, most New Yorkers could not work remotely yet live in transportation deserts. Micro-mobility has worked in other cities, but cost has put it out of reach for lower-income New Yorkers. The Equitable Commute Project is a collaboration to create ownership access to electric commute options for frontline Bronx and Brooklyn workers through a novel lending program for people without credit histories.
- **Decarbonization of buildings:** Seventy percent of NYC's greenhouse gas emissions come from 1 million buildings. The Carbon Mobilization Act, landmark legislation passed in 2019, offers both sticks (tax) and carrots (access to low-cost capital, called the Property Assessed Clean Energy Program, or PACE) to decarbonize buildings. Invest NYC SDG worked with the mayor's office to support a data tool to learn who owns and who

BOX 9-1 Continued

holds the mortgages of high-carbon buildings. The effort will target not only property owners, but also mortgage lenders to increase the use of PACE. NYC SDG is working to develop bank leadership so that banks will proactively accept PACE loans and themselves become PACE lenders.

References

Invest NYC SDG Initiative website, https://www.stern.nyu.edu/experience-stern/about/departments-centers-initiatives/centers-of-research/center-sustainable-business/research/invest-nyc-sdg-initiative.
Koval, M. 2022. Workshop Presentation, May 17, 2022.
Koval, M. 2021. Creating a Sustainable, Inclusive, and Resilient Future Economy in New York City. https://wagner.nyu.edu/files/nyc2025/Koval_NYC%20Sustainable%20Inclusive%20and%20Resilient%20Goals_NYC2025_101921.pdf.
McKibben. B. 2017. Drawdown: The Most Comprehensive Plan Ever Proposed to Reverse Global Warming. New York, NY: Penguin Books.
The Equitable Commute Project, https://equitablecommuteproject.carrd.co.

TABLE 9-1 Landscape of Financing Instruments for Capital to Make Impact

	Debt	Private Equity and VC	Other
Built Environment Sustainable Mobility	Tax Exempt Bonds, e.g., Betances Residences Green Bonds, e.g., MTA Green Bond	Accelerators e.g. The Urban Future Lab New Business Models, e.g., Carma; Bay Area Rapid Transit (BART)	Government subsidies/Philanthropic Grants Capital Leasing; Public-Private Partnerships; Grants
Renewable Energy	Asset-Backed Securities; Property Assessed Clean Energy (PACE); Green Bonds	Power Purchase Agreements; Tax Equity Financing; Real Estate Investment Trusts (REITs)	Energy Services Agreements; Energy Performance Contracts
Waste	Variable Payment Obligations, e.g., ReFED	Venture Capital, Project Finance, Private Equity, e.g., Closed Loop Partners	Government subsidies/Philanthropic Grants and "Venture Philanthropy," e.g., SEACEF
Food and Health	Microloans, e.g., Farm Service Agency (FSA)	Venture Capital, e.g., Pod Food	Crowd Funding e.g., Brooklyn Grange
Climate Resilience	Environmental Impact Bond e.g., DC EIB	Venture Capital, e.g., Quantified Ventures	Catastrophe Bond e.g., FEMA Catastrophe Bond

SOURCE: NYU, 2021. Reprinted with permission from the NYC Stern Center for Sustainable Business.

BOX 9-2
Financing Sustainable Development: Benchmarking as a Tool

Benchmarks equip investors, governments, civil society, individuals, and companies with the information they need to engage and step up to achieve transformational change (Muusse, 2022). The World Benchmarking Alliance (WBA) was launched in 2018 based on the belief that the private sector can contribute to and benefit from the global ambition of the Sustainable Development Goals (SDGs), and that corporate performance benchmarks are powerful levers for change. WBA has identified seven systems that offer a strategic framework for achieving the SDGs: food and agriculture, decarbonization and energy, urban, digital, and nature and biodiversity, with social at the core and the financial system essential to unlock change.

Keystone Companies

WBA assesses, measures, and ranks 2,000 of the world's most influential companies on their contributions to the SDGs and other internationally accepted norms using free, publicly available data. They are called "keystone companies" because of their disproportionate influence on the structure or functioning of the system in which they operate. Profiles and analyses of these companies, called SDG2000, are available on the WBA website. Among the 2,000 companies are 400 financial institutions. They include banks, insurance companies, asset managers, and asset owners (e.g., pension funds, sovereign wealth funds, and development finance institutions). Financial institutions have a dual role—being benchmarked themselves through the WBA Financial System Benchmark and mobilizing companies in the other systems through their financing activities.

Financial Benchmarking

The financial institutions are assessed on their readiness to address global sustainability transitions and their contributions to the 2030 Agenda. The WBA benchmark has three broad measurement areas: governance and strategy, respecting planetary boundaries, and adhering to societal conventions, with a total of 32 issues and indicators. This benchmark, expected for completion later in 2022, is a "race to the top" to illustrate how these 400 companies compare against one another and to understand the gaps across the areas.

Investor Space

Financial institutions also play a key role across the economy as investors, lenders, insurers, and other roles. Benchmarks empower them in their provision of analysis, identification of sustainability risks and opportunities, and assessment of company performance, as well as facilitate their engagement across sectors and allocation of capital. Twenty-seven "investor allies" are working with WBA through active ownership, knowledge sharing, market shaping/signaling, and investment research and/or fund creation. Investors are also using company scorecards to inform one-on-one engagement. Others are replicating the WBA methodology and systems lens in their own work.

Collective Impact Coalitions

WBA provides a space for allies to come together and develop coordinated action across sectors, builds upon existing efforts to raise the bar for all, mobilizes

---

BOX 9-2 Continued

action through collaboration, and identifies key issues from the benchmarking data that are catalytic to drive systemic transformation. Recent issues that have been the focus of these coalitions include ethical artificial intelligence in the digital space and a just energy transition.

References

Muusse, L. 2022. Workshop Presentation, May 18, 2022.
World Benchmarking Alliance website, https://www.worldbenchmarkingalliance.org.

SASB and other frameworks previously in competition with one another and individual corporate structures. From the Value Reporting Foundation:

> In an era where the impacts of global pandemic, climate change and growing inequality are intensifying, the concepts of sustainability and intangible value have grown in importance. Capital markets must evolve to deliver long-term value to shareholders while also helping secure the future of our people and our planet—improving reporting is an important means to this end (Medress, 2022; Value Reporting Foundation. 2020).

KEY RESEARCH PRIORITIES FOR FINANCING TO ACHIEVE THE SDGS

The committee proposes the following areas for further research to operationalize sustainable development in the areas of financing for sustainable development:

- Explore place-based initiatives in need of private investment, such as community-supported initiatives, or other means of providing capital for P3, such as those identified through the Invest NYC SDG initiative (Koval, 2022; NYU, 2021). Adam (2022) also raised the issue of near-shoring and localized value chains enabled by digitalization, the mission of the Digital Economy for Africa initiative (The World Bank, 2022a).
- Examine key ways to unlock financing for the SDGs, including local initiatives to sufficiently scale or tranche themselves to meet investor demand, and whether barriers such as debt limits, reporting requirements, and jurisdictions limit this scale.
- Identify brokers needed to "matchmake" the capital investment required to accelerate projects that will advance the SDGs, as well as to identify entities in the public, private, and nonprofit sectors with funding needs.
- Develop strategies to advance adoption of emerging integrated reporting standards that help define "stakeholder value" as opposed to shareholder

returns and deter "greenwashing," and monitor whether aligned companies and investors outperform non-ESG-aligned portfolios over time.

- Explore costs, benefits, challenges, and opportunities relating to certification standards such as Climate Bonds Certification, Green Bond Principles (as described in Chapter 6), and SDG Bonds (Dimovska, 2021). It would be useful to examine whether these standards could be integrated into social financing projects in cities (e.g., housing building projects).

POSSIBLE ACTIONABLE STEPS FOR
FINANCING TO ACHIEVE THE SDGS

The committee proposes the following possible actionable steps to operationalize sustainable development in the area of financing to achieve the SDGs:

- Public, private, and other organizations could create more blended finance options given the growing demand for positive ESG investments where social benefits are significant and measurable. With these options, more capital could be provided by local governments and new public-private or other partnerships. Impact investing, development aid, and concessionary capital could jumpstart a project or a partnership that provides access to previously underserved markets.
- National governments could provide cities and regional governments with the creditworthiness or access to capital markets so that they can issue debt to finance needed transformational projects. Risk could be redefined to account for the cost of inaction.
- Private-sector companies could participate in partnerships like the UN Global Compact (and Global Compact USA) that are setting ambitious targets for themselves and their peers, and/or transparently measure themselves against benchmarks developed by the World Benchmarking Alliance.
- Funding agencies and philanthropic organizations could promote additional investment into the development of local value chains and sustainability innovations, using a circular economy framework in the context of the COVID-19 pandemic, geopolitical conflict, and climate change.
- Governments could support research and private-sector initiatives relating to renewable energy, such as those that expand electric car charging stations, build railroads and bike lanes, and subsidize public transportation.

10

Moving Forward

This chapter compiles the committee's key research priorities and possible actionable steps needed to operationalize sustainable development by stakeholder. As stated in the committee's task (Box 1-1), these research priorities and possible actionable steps were identified mostly from presentations and discussions at the workshops held by the committee. Common areas across the eight themes discussed in this report include the need for additional data and reporting, the need for multi-stakeholder, multi-sectoral collaboration, the importance of participatory processes in decision-making, and the need for targeted financing at multiple levels from the international to the community scale. Delivering sustainability missions will require broad engagement and commitment from governments, the private sector, science funders, and civil society (ISC, 2021). The committee hopes that positive case studies provided in this report, along with the information summarized below, will guide federal and local policy makers, researchers, practitioners, civil society, educators, business and philanthropic leaders, and other stakeholders in their efforts toward sustainable development. The committee believes these recommendations are ambitious but realistic and, taken together, can make a measurable difference in a sustainable future for all.

KEY RESEARCH PRIORITIES

Education and Capacity Building (Chapter 2)	• Conduct **research investigations, case studies, and evaluations of effective efforts** building partnerships and operationalizing the SDGs at the local and subnational levels that connect to national and global levels with special focus on K-12 and university education, public outreach, and capacity building.
	• Identify **effective ways to support K-12 education initiatives** that assist students in defining, developing, and implementing their own frameworks for sustainable actions in their communities and in understanding the impacts beyond.
	• Examine issues relating to ensuring **diversity, equity, accessibility, and inclusion in K-12 STEM education** as well as leveling the playing field in access to K-12 education across school districts in the United States.
	• Examine how sustainability education programs at the undergraduate and graduate levels **can prepare all students, regardless of major, to contribute to advancing a post-2030 agenda for sustainable development,** as well as identify **best practices in field building for sustainable development at the undergraduate and graduate levels** that will be important for research and education in moving that agenda forward.
Localization of the SDGs and Indigenous Knowledge (Chapter 3)	• Understand the **synergies and tradeoffs that can help to achieve localization of the SDGs,** including the appropriate balance between economic, social, and environmental considerations at the local, national, and global levels.
	• Identify **key mechanisms that address poverty and empower vulnerable communities.**
	• Identify **governance models and arrangements** that could accelerate local transformations for sustainable development.
	• Explore **ways to make science systems more inclusive and equitable,** to involve a wider range of voices, institutions, types of knowledge, and approaches to learning **that are designed to capture local needs.**
	• Establish effective frameworks that incorporate both **conventional scientific knowledge and indigenous knowledge.**
Food Systems (Chapter 4)	• Conduct **a comprehensive analysis of the entire food system,** including production, processing, distribution, consumption, and waste disposal involving social, economic, and environmental outcomes in urban and rural areas.
	• **Examine how to transform food systems** to achieve critical progress on the SDGs and to contribute to a better future, including reducing inequalities and promoting well-being along economic, environmental, and social dimensions of sustainability.
	• **Examine the future of alternative proteins,** precision fermentation, 3D printing of meat, fish, and plant-based proteins to support sustainable, nutritious, and equitable food systems including consumer acceptance.
	• Explore the **potential impacts of targeted technological innovations** on urban and rural agriculture, agribusiness, food supply chain, animal welfare, climate change, energy, water, land use, biodiversity, health, and food loss and waste, as well as societal and cultural barriers they may encounter.

Urbanization (Chapter 5)	• Examine how to achieve **systemic transformation across multiple dimensions of the SDGs at the city-regional scale**, while addressing transnational and rural-urban linkages and externalities, including shifting burdens (social, economic, and environmental) beyond the regional borders.
	• Build a **multiscale narrative of urban change that links local, national, regional, and global activities** in the context of COVID-19, climate change, and global conflicts.
	• Improve **data collection and reporting at the local level**, including disaggregated and city-level data, as well as **create open data hubs and portals** to capture information from local agencies and community-generated data sets
	• Improve information on cities, including **by filling data gaps, especially in the Global South**, because developing cities will contribute the majority of the future urban transition in coming decades.
	• Understand the types of **data needed from cities to monitor SDG transitions** along environmental, social, *and* economic considerations. The first step would be to identify the types of data that are missing. The second step would be to determine how to collect these data across multiple locations.
	• Explore how to conduct **research on systemic equity and power** that align with research on environmental and economic transitions, because this research is often conducted in silos.
Decarbonization (Chapter 6)	• **Examine fundamental science for ocean- and nature-based carbon dioxide removal**, including chemical pathways, microbiome variability and durability of soil sequestration, forest and ocean-based proposals, and suitable reservoirs for underground or deep sea storage.
	• Conduct standards setting for **monitoring, reporting, and verification techniques** for various pathways.
	• Explore **acceptable levels of uncertainty** in certification in both technical and social dimensions including intergenerational justice.
	• Improve the understanding of possible **impacts on biodiversity, land, or ocean use** for food or other unintended consequences such as tipping points for carbon sinks becoming sources.
	• Examine technologies that enable **large-scale deployment of carbon capture, utilization, and storage**, with an emphasis on durability and scale-up. Examples include the Sleipner T carbon dioxide treatment platform and the carbon capture plant in Iceland (Panko, 2021).
Science, Technology, and Innovation for the SDGs (Chapter 7)	• **Examine the current status of achieving the SDGs in the United States** and what actions and resources are needed to advance the SDGs in the context of the economic crisis, the COVID-19 pandemic, and geopolitical conflicts.
	• **Measure success of STI partnerships**, such as the STI for SDGs roadmaps.
	• Explore **what voice cities, city networks, and their partners should have** in the multilateral system and how they can impact future international commitments.

Science and Peace (Chapter 8)	• **Strengthen SDG data hubs, partnerships, and data for SDG monitoring and enforcement** relating to science and peace and other relevant issues.
	• Explore **survey instruments on SDG 16 and the interlinkages between different variables** while supporting counties to collect data on access to justice, corruption, discrimination, and trafficking.
	• Examine how to deal with **post-conflict trauma** because global conflicts will influence future generations, including the war in Ukraine.
	• Prevent and mitigate the **effects of child soldiers** and **gender-based violence** that occurs within conflicts.
Financing to Achieve the SDGs (Chapter 9)	• **Explore place-based initiatives** in need of private investment, such as community-supported initiatives, or other means of providing capital for public-private-philanthropic partnerships.
	• Examine key ways to **unlock financing for the SDGs**, including local initiatives to sufficiently scale or tranche themselves to meet investor demand, and whether barriers such as debt limits, reporting requirements, and jurisdictions limit this scale.
	• **Identify brokers needed to "matchmake"** the capital investment required to accelerate projects that will advance the SDGs, as well as to identify entities in the public, private, and nonprofit sectors with funding needs.
	• **Develop strategies to advance adoption of emerging integrated reporting standards that help define** "stakeholder value" as opposed to shareholder returns and deter "greenwashing," and monitor whether aligned companies and investors outperform non-ESG-aligned portfolios over time.
	• Explore **costs, benefits, challenges, and opportunities relating to certification standards** such as Climate Bonds Certification, Green Bond Principles, and as SDG Bonds.

POSSIBLE ACTIONABLE STEPS BY STAKEHOLDER

The following section reorders the possible actionable steps needed to operationalize sustainable development described in each chapter by stakeholder. Although these suggestions are listed by stakeholder, their implementation would benefit from collaborative efforts by several or all stakeholders.

Federal agencies could

- Engage the public to raise awareness of the SDGs (together with education leaders). Examples include the National Climate Assemblies in many European countries, various global campaigns, and work with film makers supporting the SDGs. (Chapter 2)
- Commit to creating a Voluntary National Review (VNR) by encouraging more states and cities to conduct Voluntary Local Reviews (VLRs) and synthesize already good work at the local level to scale to a VNR roll-up. (Chapter 3)
- Provide financial incentives for local and state VLRs and consider federal and state regulatory changes to create flexibility. (Chapter 3)

- Identify strategies for carbon dioxide removal (CDR) that are place-based, community embraced, and environmentally and intergenerationally just. (Chapter 6)
- Build on the current U.S. $1 billion allocated for the four regional Direct Air Capture hubs to establish other CDR demonstration projects, such as biochar in concrete, asphalt, and soil, as well as global satellite forest monitoring. (Chapter 6)
- Set a flue point capture target of $50/tonne for hard to abate industries, including establishing an international prize competition. (Chapter 6)
- Ramp up research, development, demonstration, and deployment (RDD&D) for all forms of CDR. (Chapter 6)
- Play a leadership role in international collaboration and co-funding of research, provide international incentives for ethical deployment and scale-up, and propose an international framework for standards and monitoring, reporting, and verification (MRV) to deter national and corporate "greenwashing." (Chapter 6)
- Enhance federal coordination between agencies, the U.S. Department of Agriculture, Department of Energy, Department of Interior, Environmental Protection Agency, General Services Administration, and Department of Defense relating to incentives, MRV, siting, and accounting. (Chapter 6)
- Expand attribute-focused rather than prescriptive tax incentives for CDR, such as by focusing on sequestration durability instead of pathway-specific technology. (Chapter 6)
- Use procurement to catalyze and set standards for private-sector investments, promote incentives for sequestration not just capture, and utilize lessons learned from partnerships to engage in international dialog on ethics and environmental justice in CDR, as well as framework and standards for MRV (in collaboration with state governments and international coalitions). (Chapter 6)
- Be reoriented as the champion of SDGs through participation in a VNR. (Chapter 7)
- Provide cities and local governments with the creditworthiness or access to capital markets so that they can issue debt to finance needed transformational projects. (Chapter 9)
- Support research and private-sector initiatives relating to renewable energy, such as those that expand electric car charging stations, build railroads and bike lanes, and subsidize public transportation. (Chapter 9)

Colleges and universities could

- Undertake initiatives to assist faculty and students in developing Voluntary University Reviews (VURs) to evaluate needs and prioritization among SDGs based on an institutional mission, take actionable steps that advance progress on the SDGs at their universities, and ensure that every student regardless of major is exposed to the challenges and opportunities in sustainable development. (Chapter 2)

- Develop partnerships with local governments, universities, business communities, and civil society organizations to develop VURs to evaluate needs and take actionable steps that can advance progress toward the SDGs by their cities and local communities. (Chapter 2)
- Elevate a focus on building the field of sustainability science as a discipline to prepare the next generation for a post-2030 agenda for sustainable development. (Chapter 2)
- Help surrounding communities and cities conduct VURs and/or VLRs. (Chapter 3)
- Support training and workforce development at universities, community colleges, and tribal colleges to promote sustainable and equitable food systems, taking into account environmental, economic, and social considerations (together with the private sector). (Chapter 4)
- Create opportunities for workshop reports and journal special editions that focus sharply on identifying critical knowledge gaps relating to big data and research on cities and on producing new knowledge of special relevance to direct action, such as providing guidance to funders about areas for future work. (Chapter 5)

K-12 educational leaders could

- Initiate and support programs at the local K-12 level for students to undertake local investigations in their communities on the SDGs across diverse contexts, define and implement frameworks for sustainable development, and connect their local issues to global issues. (Chapter 2)
- Provide teachers with a network of peer mentors and a platform, such as developing and maintaining a website to host downloadable materials relating to sustainability, the SDGs, and climate change education. (Chapter 2)

Governments, the private sector, and nongovernmental organizations could

- Support initiatives that further the role of indigenous knowledge in the development of scientific knowledge. (Chapter 3)
- Discuss challenges, opportunities, and innovative strategies for sustainable and equitable urban food systems in the United States. (Chapter 4)
- Accelerate initiatives toward sustainable and equitable food systems with an appropriate sense of urgency, because the urban population is projected to increase rapidly in the coming decades. (Chapter 4)
- Work together to transform food systems in the United States and to promote supply chains that are resilient and sustainable in the short and long terms. (Chapter 4)
- Promote decarbonization in agriculture, industry, and energy production, including by building carbon-neutral cities, strengthening climate

education and engagement, and encouraging low-carbon lifestyles for mobility, housing, and consumption. (Chapter 6)

- Discuss a new vision for a sustainable and resilient future beyond the 2030 Agenda for Sustainable Development, because the world is not on track to achieve the original 2015 targets by 2030. (Chapter 7)
- Look to all sectors (not just national governments) for partnerships that consider ways to have a greater voice in shaping our multilateral system and future shared commitments. (Chapter 7)
- Become involved in the efforts at the United Nations (UN) to assist volunteer countries in developing their science, technology, and innovation (STI) for SDG roadmaps and to facilitate knowledge exchange and transfer at the local level. (Chapter 7)
- Support humanitarian efforts in conflict areas. (Chapter 7)
- Design a new global social compact that emphasizes peace building and promotes a global system that builds on care, sharing, sufficiency, and respect for human and non-human living beings, perhaps as part of the effort to negotiate the initiative that follows the SDGs beyond 2030. (Chapter 8)
- Create peer groups for implementation and monitoring the SDGs (e.g., cities learn best from other cities), dealing with crises situations, and facilitating exchange among justice actors, peace builders, and inequality experts. (Chapter 8)
- Promote positive examples for supporting Ukrainian scientists, and additional efforts to support science, engineering, and medical professionals in other nations including Afghanistan, Myanmar, South Sudan, Syria, Venezuela, Mexico, and Nicaragua. (Chapter 8)
- Conduct studies and dialogues that help to advance control of new and emerging weapon systems as has been the case with nuclear weapons. (Chapter 8)
- Create more blended finance options given the growing demand for positive environmental, social, and governance investments where social benefits are significant and measurable. (Chapter 9)

Local officials could

- Engage students from universities, community colleges, and Minority-Serving Institutions to organize student projects in cities. (Chapter 2)
- Commit their support to the SDGs and use the framework to align local policies and initiatives. (Chapter 3)
- Learn from excellent case studies of knowledge networks, which effectively incorporate indigenous knowledge for advancing sustainability. (Chapter 3)
- Enhance cross-sector collaborations and engage communities to transform urban food systems. (Chapter 4)
- Convene diverse, inclusive groups in workshops to focus on the key research priorities described in Chapter 5 in the context of the COVID-19 pandemic, climate change, and global conflicts. (Chapter 5)

International organizations could

- Establish and maintain databases for international research on urbanization. (Chapter 5)
- Maintain the continuity of experience for the UN's various scientific groups, such as the 10-Member Group and the Global Sustainable Development Report (GSDR) group, which might play an important role in the post-2030 processes. (Chapter 7)
- Convene experts who have participated in the 10-Member Group, GSDR, and other UN SDGs efforts to share knowledge and experiences with current members of scientific groups to ensure the continuity of knowledge and to engage with young people. (Chapter 7)

Private-sector companies could

- Lead the advancement of STI in sustainable design and operation toward scalable achievement of the SDGs and influence the sustainability profile of multiple economic sectors. (Chapter 7)
- Enhance public-private partnerships to support disaster recovery through STI. (Chapter 7)
- Participate in partnerships like the UN Global Compact (and Global Compact USA) that are setting ambitious targets for themselves and their peers, and/or transparently measure themselves against benchmarks developed by the World Benchmarking Alliance. (Chapter 9)

Funding agencies and philanthropic organizations could

- Learn from excellent case studies of knowledge networks, which effectively incorporate indigenous knowledge for advancing sustainability. (Chapter 3)
- Highlight and support examples of effective solutions to the SDGs and sustainability challenges at the local, national, and global levels. (Chapter 7)
- Invest in organizations between state and society that can contribute to solutions that address the increasing number of wars, conflicts, and migration of displaced people, and support scientific communities in long-term sustainable development, including rebuilding efforts. (Chapter 8)
- Promote additional investment into the development of local value chains and sustainability innovations, using a circular economy framework in the context of the COVID-19 pandemic, geopolitical conflict, and climate change. (Chapter 9)

Accelerating progress on operationalizing sustainable development involving all levels of government and all sectors of society can be a major stepping stone to realize the optimistic future envisioned by the 17 Sustainable Development Goals.

References

Adam, J-P. 2022. Remarks at the Workshop on Operationalizing Sustainable Development, May 4, 2022.

African Network of Cities. 2022. https://aston-network.org.

Apolitical. 2022a. Developing the Next Generation of SDG Leadership. https://apolitical.co/solution-articles/en/universities-cities-sustainable-development.

Apolitical. 2022b. How Universities Can Help Cities Achieve the Sustainable Development Goals. https://apolitical.co/solution-articles/en/universities-cities-sustainable-development.

APS (American Physical Society). 2022. July 2022 Newsletter. https://engage.aps.org/fps/resources/newsletters/july-2022.

Arcusa, S. 2022. Challenges to Sustainable Development during Decarbonization: Lessons Learned from Research on the Certification of Carbon Removal and Decarbonization Pathways. Presentation at the Workshop on Operationalizing Sustainable Development, May 16, 2022.

Arcusa, S., and S. Sprenkle-Hyppolite. 2022. Snapshot of the carbon dioxide removal certification and standards ecosystem (2021–2022). *Climate Policy.* https://doi.org/10.1080/14693062.2022.2094308.

Basnyat, A. 2022. Operationalizing Sustainable Development: Measuring Peace, Justice & Inclusion (SDG 16). Presentation at the Workshop on Operationalizing Sustainable Development, May 16, 2022.

Beijing Center. 2022. https://thebeijingcenter.org/course/chinese-medicine.

Bourland, C. 2022. Operationalizing Sustainable Development. Presentation at the First Meeting of the Committee on Operationalizing Sustainable Development, February 22, 2022.

Brookings. 2021. City Playbook for Advancing the SDGs: A collection of how-to briefs on advancing the Sustainable Development Goals locally. https://www.brookings.edu/multi-chapter-report/city-playbook-for-advancing-the-sdgs.

Brundiers, K., M. Barth, G. Cebrián, M. Cohen, L. Diaz, S. Doucette-Remington, W. Dripps, G. Habron, N. Harré, M. Jarchow, K. Losch, J. Michel, Y. Mochizuki, M. Rieckmann, R. Parnell, P. Walker, and M. Zint. 2020. Key competencies in sustainability in higher education—toward an agreed-upon reference framework. *Sustainability Science* 13-19. https://doi.org/10.1007/s11625-020-00838-2.

Bumb, P. 2022. Carbon Clean: Technology to Achieve 'Net Zero'. Revolutionising Industrial Decarbonization. Presentation at the Workshop on Operationalizing Sustainable Development,
May 16, 2022.

Business and Sustainable Development Commission. 2017. Better Business Better World. https://
sustainabledevelopment.un.org/content/documents/2399BetterBusinessBetterWorld.pdf.

Bürgerrat. 2022. Climate Assembly in Austria. https://www.buergerrat.de/en/news/climate-assembly-
in-austria.

Burns, E. 2022. Presentation at the Workshop on Operationalizing Sustainable Development, May 16, 2022.

C40. 2022. https://www.c40.org

Campbell, D. E. 2006. What is education's impact on civic and social engagement? Measuring
the Effects of Education on Health and Civic Engagement: Proceedings of The Copenhagen
Symposium. Available at https://www.oecd.org/education/innovation-education/37425694.pdf.

Carbon180, https://carbon180.org.

Casacio, C. A., L. S. Madsen, A. Terrasson, M. Waleed, K. Barnscheidt, B. Hage, M. A. Taylor,
and W. P. Bowen. 2021. Quantum-enhanced nonlinear microscopy. *Nature* 594. https://doi.
org/10.1038/s41586-021-03528-w.

Center for Sustainable Systems, University of Michigan. 2022. U.S. Cities Factsheet. https://css.
umich.edu/publications/factsheets/built-environment/us-cities-factsheet.

Ciambra, A., and R. Martinez. 2022. Voluntary Local Reviews, VLRs toolbox: From data analysis to
citizen engagement when monitoring the SDGs. Swedish International Centre for Local Democracy.
https://unhabitat.org/sites/default/files/2022/03/toolbox-for-voluntary-local-reviews-vlr.pdf.

Circular Copenhagen. 2022. Resource and Waste Management Plan. 2022. https://circularcph.cph-
solutionslab.dk/cc/home.

City of Los Angeles. 2022. Biodiversity in Los Angeles Index. https://sdg.lamayor.org/our-work/
projects/biodiversity-los-angeles.

Climate Bonds Certification. 2022. https://www.climatebonds.net/certification.

CMU (Carnegie Mellon University). 2022. The Sustainability Initiative at CMU: Toward a more sustainable future—for everyone. https://www.cmu.edu/leadership/the-provost/provost-priorities/
sustainability-initiative/index.html.

CMU. 2021. CMU Voluntary University Review of the Sustainability Development Goals 2021. https://
local2030.org/library/848/Carnegie-Mellon-University-Voluntary-University-Review-2021.pdf.

CMU. 2020. CMU Voluntary University Review of the Sustainability Development Goals 2020.
https://www.cmu.edu/leadership/the-provost/provost-priorities/sustainability-initiative/cmu-vur-
2020.pdf.

Colglazier, E. W. 2018. The Sustainable Development Goals: Roadmaps to progress. *Science Diplomacy* 7(1). https://www.sciencediplomacy.org/editorial/2018/sdg-roadmaps.

Connors, C. 2022. Hawai'i Local 2030 Hub. Presentation at the Workshop on Operationalizing Sustainable Development, April 21, 2022.

Crippa, M., E. Solazzo, D. Guizzardi, et al. 2021. Food systems are responsible for a third of
global anthropogenic GHG emissions. *NatureFood* 2:198–209. https://doi.org/10.1038/
s43016-021-00225-9.

Dimovska, M. 2021. SDG Bonds: A contribution to Building Forward Better. https://www.undp.org/
eurasia/blog/sdg-bonds-contribution-building-forward-better.

DOE (U.S. Department of Energy). 2022. Four Regional Clean Direct Air Capture Hubs: Bipartisan
Infrastructure Law. https://www.energy.gov/bil/four-regional-clean-direct-air-capture-hubs.

Draper, K. 2022. Biochar: Safe, Scalable, and Shovel-Ready. Presentation at the Workshop on
Operationalizing Sustainable Development, May 16, 2022.

EC. (European Commission). 2019. The European Green Deal. https://ec.europa.eu/info/sites/default/
files/european-green-deal-communication_en.pdf.

Ellen MacArthur Foundation. 2022. A Circular Economy for Food Will Help People and Nature
Thrive. https://ellenmacarthurfoundation.org/topics/food/overview.

Essential Costa Rica. 2022. Costa Rica Takes Sustainable Travel to the Next Level. https://www.visit costarica.com/en/costa-rica/blog/costa-rica-takes-sustainable-travel-next-level#:~:text= Finca%20Rosa%20Blanca&text=As%20one%20of%20the%20original,directly%20benefit%20 the%20surrounding%20community.

EU (European Union). 2022. New European Bauhaus. https://europa.eu/new-european-bauhaus/ index_en.

GCSA (Group of Chief Scientific Advisors). 2021. A Systemic Approach to the Energy Transition in Europe: Scientific Advice to Strengthen the Resilience of the European Energy Sector, Scientific Opinion of the GCSA to the European Commission. https://op.europa.eu/en/publication-detail/-/ publication/d01f956f-de07-11eb-895a-01aa75ed71a1/language-en.

GEA. 2012. *Global Energy Assessment - Toward a Sustainable Future*. Cambridge, UK and New York, NY: Cambridge University Press

Global Carbon Removal Partnership. 2022. https://carbonremovalpartnership.net.

Global Island Partnership. 2022. https://www.glispa.org.

Global Tapestry of Alternatives. 2022. https://globaltapestryofalternatives.org.

Global Urban Development websites, www.globalurban.net; www.globalurban.org.

Gordon, A. R. 2022. The UN Global Compact and Financing the SDGs. Presentation at the Workshop on Operationalizing Sustainable Development, May 17, 2022.

Graeber, D., and D. Wengrow. 2022. *The Dawn of Everything: A New History of Humanity*. New York, NY: Farrar, Straus and Giroux. https://us.macmillan.com/books/9780374157357/ thedawnofeverything.

Grantham, A. 2022. Operationalizing Sustainable Development: Local Strategies. Sustainable and Equitable Food Systems. Presentation at the Workshop on Operationalizing Sustainable Development, April 18, 2022.

Green Bond Principles. 2022. https://www.icmagroup.org/sustainable-finance/the-principles-guidelines- and-handbooks/green-bond-principles-gbp.

Häfele, W., J. Anderer, A. McDonald, and N. Nakicenovic. 1981. *Energy in a Finite World: Paths to a Sustainable Future*, Volume 1. Cambridge, MA: Ballinger. https://pure.iiasa.ac.at/id/eprint/1539.

Hawai'i Green Growth website, https://www.hawaiigreengrowth.org.

Hawai'i's Open-Data Aloha+ Challenge Dashboard. 2022. https://aloha-challenge.hawaiigreen growth.org/dashboard.

IAP (The InterAcademy Partnership) website, https://www.interacademies.org/iap/about.

IAP (InterAcademy Partnership). 2022. Join the Advocacy Campaign to Support At-Risk, Displaced and Refugee Scientists by Signing the Science in Exile Declaration. https://www.interacademies.org/ news/international-science-organizations-unite-support-risk-displaced-and-refugee-scientists.

IFAD (International Fund for Agricultural Development). 2019. Partnering with Indigenous Peoples for the SDGs. https://www.ifad.org/documents/38714170/41390728/policybrief_indigenous_ sdg.pdf/e294b690-b26c-994c-550c-076d15190100.

IIE (Institute of International Education). 2022. Scholar Rescue Fund. https://www.scholarrescuefund.org.

IPCC (Intergovernmental Panel on Climate Change). 2022. Climate Change 2022: Mitigation of Climate Change. Working Group III contribution to the Sixth Assessment Report of the IPCC. https://report.ipcc.ch/ar6wg3/pdf/IPCC_AR6_WGIII_FinalDraft_FullReport.pdf.

IPCC. 2018. Global Warming of 1.5°C. An IPCC Special Report on the impacts of global warming of 1.5°C above pre-industrial levels and related global greenhouse gas emission pathways, in the context of strengthening the global response to the threat of climate change, sustainable development, and efforts to eradicate poverty.

ISC (International Science Council). 2021. Unleashing Science: Delivering Missions for Sustain- ability. doi: 10.24948/2021.04.

ISC. 2020. A Synthesis of Research Gaps: For science to enable societies to accomplish the Sus- tainable Development Goals by 2030. https://council.science/wp-content/uploads/2020/06/ 202109_A-Synthesis-of-Research-Gaps_Final.pdf.

Invest NYC SDG Initiative website, https://www.stern.nyu.edu/experience-stern/about/de-partments-centers-initiatives/centers-of-research/center-sustainable-business/research/invest-nyc-sdg-initiative.

Jessen, T. D., N. C. Ban, N. X. Claxton, and C. T. Darimont. 2021. Contributions of indigenous knowledge to ecological and evolutionary understanding. *Frontiers in Ecology and the Environment* 20(2):93-101. https://doi.org/10.1002/fee.2435.

Judge, P. 2022. Alphabet, Microsoft, and Salesforce Launch $500m First Movers Carbon Capture Group. https://www.datacenterdynamics.com/en/news/alphabet-microsoft-and-salesforce-launch-500m-first-movers-carbon-capture-group.

Kanbur, R. 2022. Remarks at the Workshop on Operationalizing Sustainable Development, April 18, 2022.

Kothari, A. 2022. Presentation at the Workshop on Operationalizing Sustainable Development, April 2022.

Kothari, A., A. Salleh, A. Escobar, F. Demaria, and A. Acosta. 2019. Pluriverse: A Post-Development Dictionary. Tulika Books and Authors Upfront, Delhi.

Koval, M. 2022. Invest NYC SDG. Presentation at the Workshop on Operationalizing Sustainable Development, May 17, 2022.

Koval, M. 2021. Creating a Sustainable, Inclusive, and Resilient Future Economy in New York City. https://wagner.nyu.edu/files/nyc2025/Koval_NYC%20Sustainable%20Inclusive%20and%20Resilient%20Goals_NYC2025_101921.pdf.

Leonardsen, L. 2022. Presentation at the Workshop on Operationalizing Sustainable Development, April 21, 2022.

Lilja, J. 2022. Remarks at the Workshop on Operationalizing Sustainable Development, May 16, 2022.

Local2030 Islands Network website, https://www.islands2030.org.

Locke, R. 2022. Operationalizing Sustainable Development: Global Strategies and the Science of Peace. Presentation at the Workshop on Operationalizing Sustainable Development, May 16, 2022.

Mashelkar, R. A. 2001. Intellectual property rights and the Third World. *Current Science* 81(8): 955–965. https://www.jstor.org/stable/24106520.

McCaffrey, M. 2022. Education & Capacity Building: Global Strategies Presentation at the Workshop on Operationalizing Sustainable Development, April 18, 2022.

McKibben, B. 2017. *Drawdown: The Most Comprehensive Plan Ever Proposed to Reverse Global Warming.* New York, NY: Penguin Books.

Medress, A. 2020. Answering Your Questions About the Value Reporting Foundation. https://www.sasb.org/blog/answering-your-questions-about-the-value-reporting-foundation.

Mendelson, S. 2022a. Heinz in the Hague. https://www.heinz.cmu.edu/media/2022/July/heinz-in-the-hague.

Mendelson, S. 2022b. Remarks at the Workshop on Operationalizing Sustainable Development, April 18, 2022.

Milan Urban Food Policy Pact. 2022. https://www.milanurbanfoodpolicypact.org.

Morning Consult and United Nations Foundation. 2021. Sustainable Development Goals: Awareness, Priorities, Impact on Business. https://s3.amazonaws.com/media.unfoundation.org/2021/12/MorningConsult_UN-SDG-Presentation-Deck-D3-MPR_JJM_CB-.pdf.

Mueller, J. 2022. Operationalizing Sustainable Development and Science and Technology Cooperation: Global Strategies–Framing Remarks. Presentation at the Workshop on Operationalizing Sustainable Development, May 4, 2022.

Müller, E. 2022. Presentation at the Workshop on Operationalizing Sustainable Development, April 18, 2022.

Muusse, L. 2022. Financing for Sustainable Development: The Work of the World Benchmarking Alliance. Presentation at the Workshop on Operationalizing Sustainable Development, May, 2022.

Nakicenovic, N. 2022. Perspectives on the pervasive energy-systems transformations. *Oxford Open Energy* 1:1-3. https://doi.org/10.1093/ooenergy/oiab005.

Nakicenovic, N., and P. D. Lund. 2021. Could Europe become the first climate-neutral continent? *Nature* 596(7873):486. doi: 10.1038/d41586-021-02311-1.

NASEM (National Academies of Sciences, Engineering, and Medicine). 2022a. Breakthrough Prize Foundation Partners with U.S. National Academy of Sciences to Support Scientists Forced to Flee Ukraine. https://www.nationalacademies.org/news/2022/03/breakthrough-prize-foundation-partners-with-u-s-national-academy-of-sciences-to-support-scientists-forced-to-flee-ukraine.

NASEM. 2022b. *Data-Informed Societies Achieving Sustainability: Tasks for the Global Scientific, Engineering, and Medical Communities: Proceedings of a Workshop–in Brief.* Washington, DC: The National Academies Press. https://doi.org/10.17226/26513.

NASEM. 2022c. Safe Passage for Scientists: Evacuating Scientists and Engineers from Afghanistan https://www.nationalacademies.org/news/2022/02/safe-passage-for-scientists-evacuating-scientists-and-engineers-from-afghanistan.

NASEM. 2021a. *2021 Nobel Prize Summit: Our Planet, Our Future: Proceedings of a Summit.* Washington, DC: The National Academies Press. https://doi.org/10.17226/26310.

NASEM. 2021b. *Progress, Challenges, and Opportunities for Sustainability Science: Proceedings of a Workshop–in Brief.* Washington, DC: The National Academies Press. https://doi.org/10.17226/26104.

NASEM. 2020. *Strengthening Sustainability Programs and Curricula at the Undergraduate and Graduate Levels.* Washington, DC: The National Academies Press. https://doi.org/10.17226/25821.

NASEM. 2019. *Reducing Impacts of Food Loss and Waste: Proceedings of a Workshop.* Washington, DC: The National Academies Press. https://doi.org/10.17226/25396.

National Center for Complementary and Alternative Medicine (NCCAM), National Institutes of Health (NIH). 2011. https://www.nih.gov/news-events/news-releases/nih-launches-web-resource-complementary-alternative-medicine.

Nixon, J. H., and M. A. Weiss. 2010. Sustainable Economic Development Strategies. Washington, DC: Global Urban Development. https://www.globalurban.org/Sustainable_Economic_Development_Strategies.pdf.

NYU (New York University). 2022. Pathfinders for Peaceful, Just, and Inclusive Societies. https://cic.nyu.edu/programs/sdg16plus.

NYU. 2021. Invest NYC SDG: A Finance White Paper: Models for Financing the UN Sustainable Development Goals. NYU Stern Center for Sustainable Business. https://www.stern.nyu.edu/sites/default/files/assets/documents/Invest%20NYC%20SDG%20Finance%20White%20Paper%203.12.21.pdf.

OECD (Organisation for Economic Co-operation and Development). 2021a. Blended Finance. https://www.oecd.org/dac/financing-sustainable-development/blended-finance-principles.

OECD. 2021b. Why Local? Why Now? Strengthening Intermediary Cities to Achieve the SDGs. https://oecd-development-matters.org/2021/07/06/why-local-why-now-strengthening-intermediary-cities-to-achieve-the-sdgs.

O'Donnell, C. 2022. Operationalizing Sustainable Development at the Local Level through K-12 Education. Presentation at the Workshop on Operationalizing Sustainable Development, April 18, 2022.

O'Donnell, C. 2018. Science Education, Identify, and Civic Engagement: Empowering Youth through the UN Sustainable Development Goals. G7 Executive Talk Series.

O'Hara, S. 2022. Restorative Urban Agriculture. Presentation at the Workshop on Operationalizing Sustainable Development, April 18, 2022.

O'Hara, S. 2018. The Five Pillars of Economic Development: A Study of a Sustainable Future for Ward 7 and 8 in Washington, D.C. Available at https://docs.udc.edu/causes/Five-Pillars-DC-Final-05-2018.pdf.

O'Hara, S. 2017. The urban food hubs solution: Building capacity in urban communities. *Metropolitan Universities Journal* 28(1) (Winter 2017).

O'Hara, S., and E. C. Toussaint. 2021. Food access in crisis: Food security and COVID-19. *Ecological Economics* 180:106859. https://doi.org/10.1016/j.ecolecon.2020.106859.

Open Government Partnership. 2022. Strengthen the Population's Engagement for Monitoring the Implementation of the Municipal Agenda 2030. (BRSP0002). https://www.opengovpartnership. org/members/sao-paulo-brazil/commitments/BRSP0002.

Otto, I. M., J. F. Donges, R. Cremades, A. Bhowmik, R. J. Hewitt, W. Lucht, J. Rockström, F. Allerberger, M. McCaffrey, S. S. P. Doe, A. Lenferna, N. Morán, D. P. van Vuuren, and H. J. Schellnhuber. 2020. Social tipping dynamics for stabilizing Earth's climate by 2050. *Proceedings of the National Academy of Sciences of the United States of America* 117(5):2354–2365. https://www.pnas. org/doi/10.1073/pnas.1900577117.

Pahnke, J., C. O'Donnell, and M. Bascope, 2019. Using Science to Do Social Good: STEM Education for Sustainable Development. Position paper at the second International Dialogue on STEM Education, Berlin, Germany.

Panko. 2021. World's Largest Carbon Capture Plant Opens in Iceland. *Smithsonian Magazine*. https://www.smithsonianmag.com/smart-news/worlds-largest-carbon-capture-plant-opens-iceland-180978620.

Parnell, S. 2022. Remarks at the Workshop on Operationalizing Sustainable Development, May 2022.

Pipa, T. 2022. Localization of the SDGs. Presentation at the Workshop on Operationalizing Sustainable Development, April 21, 2022.

Pipa, T., K. Rasmussen, and K. Pendrak. 2022. The State of the Sustainable Development Goals in the United States. American Leadership on the Sustainable Development Goals. https://www. brookings.edu/wp-content/uploads/2022/03/2022_Brookings_State-of-SDGs-in-the-US.pdf.

Radical Ecological Democracy website, www.radicalecologicaldemocracy.org.

Raworth, K. 2017. *Doughnut Economics: How to Think Like a 21st Century Economist*. White River Junction, VT: Chelsea Green Publishing.

REN 21. 2022. Renewables 2022 Global Status Report. (Paris: REN21 Secretariat). https://www. ren21.net/wp-content/uploads/2019/05/GSR2022_Full_Report.pdf.

Rodale Institute. 2020. Regenerative Organic Agriculture and the Soil Carbon Solution. https:// rodaleinstitute.org/education/resources/regenerative-agriculture-and-the-soil-carbon-solution.

Sachs, J. D., G. Lafortune, C. Kroll, G. Fuller, and F. Woelm. 2022. Sustainable Development Report 2022. From Crisis to Sustainable Development: the SDGs as Roadmap to 2030 and Beyond. Cambridge: Cambridge University Press. https://s3.amazonaws.com/sustainabledevelopment. report/2022/2022-sustainable-development-report.pdf.

Sachs, J. D., G. Schmidt-Traub, M. Mazzucato, D. Messner, N. Nakicenovic, and J. Rockström. 2019. Six Transformations to achieve the Sustainable Development Goals. *Nature Sustainability* 2:805-814.

Saiz, E. 2022. Remarks at the Workshop on Operationalizing Sustainable Development, April 21, 2022.

SAPEA (Science Advice for Policy by European Academies). 2021. A Systemic Approach to the Energy Transition in Europe. https://doi.org/10.26356/energytransition.

Schueman, L. J. 2021. A Healthy Economy Should be Designed to Thrive, Not Grow. https://www.one earth.org/a-healthy-economy-should-be-designed-to-thrive-not-grow.

Semida, W. M., H. R. Beheiry, M. Sétamou, C. R. Simpson, T. A. A. El-Mageed, M. M. Rady, and S. D. Nelson. 2019. Biochar implications for sustainable agriculture and environment: A review. *South African Journal of Botany* 127:333-347. https://doi.org/10.1016/j.sajb.2019.11.015.

17 Rooms website. 2022. https://www.brookings.edu/project/17-rooms.

Slowinski, R. 2022. Science Community for Ukraine. Presentation at the Workshop on Operationalizing Sustainable Development, May 16, 2022.

Smithsonian Science Education Center website, https://ssec.si.edu.

Stohl, R. 2018. Child Soldiers Pose a National Security Threat. https://www.stimson.org/2018/ child-soldiers-pose-national-security-threat.

Stuiver, M., and S. O'Hara. Food Connects Washington DC in 2050—A vision for urban food aystems as the centerpieces of a circular economy. *Sustainability* 13(14):7821. https://doi.org/10.3390/su13147821.

Tare, M. 2022. Education and Capacity Building. Presentation at the Workshop on Operationalizing Sustainable Development, April 18, 2022.

The Equitable Commute Project, https://equitablecommuteproject.carrd.co.

Tilmes, K. 2022. Remarks at the Workshop on Operationalizing Sustainable Development, May 4, 2022.

Tippet, B. 2020. Paying for the Pandemic and a Just Transition. https://longreads.tni.org/wp-content/uploads/2020/11/Paying-for-the-Panemic-and-a-Just-Transition-TNI-1.pdf.

Truman Center. 2022. Broadening Diplomatic Engagement across America: Report of the Truman Center City & State Diplomacy Task Force. https://www.trumancenter.org/issues/subnational-diplomacy.

TWI2050 (The World in 2050). 2020. Innovations for Sustainability. Pathways to an Efficient and Post-Pandemic Future. Report prepared by The World in 2050 initiative. International Institute for Applied Systems Analysis (IIASA), Laxenburg, Austria. http://pure.iiasa.ac.at/id/eprint/16533.

TWI2050. 2019. The Digital Revolution and Sustainable Development: Opportunities and Challenges. Report prepared by The World in 2050 initiative. IIASA, Laxenburg, Austria. http://pure.iiasa.ac.at/id/eprint/15913.

TWI2050. 2018. Transformations to Achieve the Sustainable Development Goals. http://pure.iiasa.ac.at/15347.

UCLG Learning. 2022. https://learning.uclg.org.

UN (United Nations). 2022a. Goal 10: Reduce inequality within and among countries. https://www.un.org/sustainabledevelopment/inequality.

UN. 2022b. GSDR (Global Sustainable Development Report) 2023. https://sdgs.un.org/gsdr/gsdr2023.

UN. 2022c. Operationalizing Leaving No One Behind: Good Practice Note for UN Country Teams. https://unsdg.un.org/sites/default/files/2022-04/Operationalizing%20LNOB%20-%20final%20with%20Annexes%20090422.pdf.

UN. 2022d. Technology Facilitation Mechanism (TFM). https://sdgs.un.org/tfm.

UN. 2022e. Voluntary National Reviews: Synthesis of Main Messages: 2022 High-Level Political Forum on Sustainable Development, Secretariat Background Note. https://hlpf.un.org/sites/default/files/inline-files/Synthesis%20of%20VNR%20main%20messages%204%20July.pdf.

UN. 2021a. Guidebook for the Preparation of Science, Technology and Innovation for SDGs Roadmaps. https://sdgs.un.org/sites/default/files/2021-06/GUIDEBOOK_COMPLETE_V03.pdf.

UN. 2021b. Progress Report of the Global Pilot Programme on STI for SDGs Roadmaps. https://sdgs.un.org/sites/default/files/2021-04/Progress%20Report%20of%20Global%20Pilot%20Programme%20of%20STI%20Roadmaps_2021_1.pdf.

UN. 2021c. The State of Food Security and Nutrition in the World. https://www.fao.org/publications/sofi/2021/en/.

UN. 2020. Science, Technology and Innovation (STI) for SDGs Roadmaps Progress Report: The Global Pilot Programme on STI for SDGs Roadmaps. https://sustainabledevelopment.un.org/content/documents/269413_Pilot_Progress_Report_full_version_July_2020.pdf.

UN. 2019a. 2019 Revision of World Population Prospects. https://population.un.org/wpp.

UN. 2019b. The Future Is Now: Science for Achieving Sustainable Development. Global Sustainable Development Report 2019. https://sdgs.un.org/sites/default/files/2020-07/24797GSDR_report_2019.pdf.

UN. 2016. Roadmap for Localizing the SDGs: Implementation and Monitoring at the Subnational Level. https://sustainabledevelopment.un.org/content/documents/commitments/818_11195_commitment_ROADMAP%20LOCALIZING%20SDGS.pdf.

UNDP (United Nations Development Programme). 2022a. SDG16 Survey Initiative: Implementation Manual 2022, VERSION 1.0. https://www.unodc.org/documents/data-and-analysis/sdgs/SDG16_Survey_Initiative_-_Implementation_Manual.pdf.

UNDP. 2022b. The SDGs in Action. https://www.undp.org/sustainable-development-goals.

UNDP. 2022c. The Sustainable Development Goals Report 2022. https://unstats.un.org/sdgs/report/2022/The-Sustainable-Development-Goals-Report-2022.pdf.

UNDP. 2020. Human Development Report 2020: The Next Frontier: Human Development and the Anthropocene. https://hdr.undp.org/system/files/documents//hdr2020pdf.pdf.

UNDP, UNODC (United Nations Office on Drugs and Crime), and OHCHR (Office of the United Nations High Commissioner for Human Rights). 2022. SDG16 Survey Initiative: Implementation Manual. https://www.undp.org/publications/sdg16-survey-initiative-implementation-manual.

UNESCO (United Nations Educational, Scientific and Cultural Organization). 2021. Engineering for Sustainable Development. https://unesdoc.unesco.org/ark:/48223/pf0000375644.locale=en.

UNHCR (United Nations High Commissioner for Refugees). 2022. Refugee Statistics. https://www.unrefugees.org/refugee-facts/statistics/#:~:text=26.6%20million%20refugees%20in%20the,4.4%20million%20asylum-seeker.

UNICEF (United Nations Children's Fund). 2021. Children Recruited by Armed Forces or Armed Groups. https://www.unicef.org/protection/children-recruited-by-armed-forces.

UN Statistics Division. 2022. Praira Group on Governance Statistics. https://unstats.un.org/unsd/methodology/citygroups/praia.cshtml. And re-alpha/place right before UNODC ref.

UNODC (United Nations Office on Drugs and Crime). 2021. Monitoring SDG 16: Key Figures and Trends. https://www.unodc.org/documents/data-and-analysis/statistics/DataMatters2_sdg16.pdf.

United Nations Foundation. 2022. Library of U.S. SDG Resource and Tools. https://s3.amazonaws.com/media.unfoundation.org/2022/07/Resource-Guide-to-US-SDGs-Updated-July-2022.pdf.

Urban Policy Platform. 2022. https://urbanpolicyplatform.org/urban-rural-linkages.

USAID (U.S. Agency for International Development). 2018. Cities Development Initiative. https://www.usaid.gov/philippines/partnership-growth-pfg/cdi.

USDA (U.S. Department of Agriculture). 2022. Food Waste FAQs. https://www.usda.gov/foodwaste/faqs.

USDA. 2021. Food Security Status of U.S. Households in 2020. https://www.ers.usda.gov/topics/food-nutrition-assistance/food-security-in-the-u-s/key-statistics-graphics/#:~:text=In%202020%3A,with%20adults%2C%20were%20food%20insecure.

USDA. 2015. NAL Agricultural Thesaurus. https://agclass.nal.usda.gov/vocabularies/nalt/concept?uri=https://lod.nal.usda.gov/nalt/142350.

USFS (U.S. Forest Service). 2022. Community Forest Program. https://www.fs.usda.gov/managing-land/private-land/community-forest.

Value Reporting Foundation. 2020. Communications Toolkit, November 2020. https://www.goodgovernance.academy/wp-content/uploads/2020/04/Value-Reporting-Foundation-Communications-Toolkit.pdf.

Vikalp Sangam. 2022. https://vikalpsangam.org.

Vorisek, D., and S. Yu. 2020. Understanding the Cost of Achieving the Sustainable Development Goals. Policy Research Working Paper 9146. https://documents1.worldbank.org/curated/en/744701582827333101/pdf/Understanding-the-Cost-of-Achieving-the-Sustainable-Development-Goals.pdf.

Watkins, A. 2018. Takeaways and Policy Recommendations: Global Solutions Summit 2018. https://www.globalsolutionssummit.com/uploads/3/1/5/5/31554571/takeaways_and_policy_recommendations_—_final.pdf.

WCED (World Commission on Environment and Development). 1987. *Our Common Future*. Oxford: Oxford University Press.

Weiss, M. A. 2022. Sustainable Innovation and Inclusive Prosperity: Porto Alegre Sustainable Innovation Zone (ZISPOA). Presentation at the Workshop on Operationalizing Sustainable Development, April 21, 2022.

Weiss, M. A. 2019. The Porto Alegre Sustainable Innovation Zone (ZISPOA). https://www.globalurban.org/ZISPOA_description_and_bibliography.pdf.

Weiss, M. A., and J. H. Nixon. 2011. The Global Future of Green Capitalism. Washington, DC: Global Urban Development. https://www.globalurban.org/Green_Capitalism.pdf.

White House. 2022. Critical and Emerging Technologies List Update. A Report by the Fast Track Action Subcommittee on Critical and Emerging Technologies of the National Science and Technology Council. https://www.whitehouse.gov/wp-content/uploads/2022/02/02-2022-Critical-and-Emerging-Technologies-List-Update.pdf.

Wiek, A., L. Withycombe, and C. L. Redman. 2011. Key competencies in sustainability: A reference framework for academic program development. *Sustainability Science* 6:203-218. https://doi.org/10.1007/s11625-011-0132-6.

World Bank. 2022a. The Digital Economy for Africa Initiative. https://www.worldbank.org/en/programs/all-africa-digital-transformation.

World Bank. 2022b. Water in Agriculture. https://www.worldbank.org/en/topic/water-in-agriculture#1.

World Benchmarking Alliance website, https://www.worldbenchmarkingalliance.org.

World Central Kitchen. 2022. Fearing for Safety, Thousands of Families Flee Their Homes in Ukraine. https://wck.org/relief/activation-chefs-for-ukraine.

World Economic Forum. 2019. What Does the World Really Think About the Sustainable Development Goals? https://www.weforum.org/agenda/2019/09/un-sustainable-development-goals.

World Justice Project. 2022. Acceleration Actions for Peace, Justice, and Strong Institutions. https://worldjusticeproject.org/our-work/engagement/acceleration-actions-peace-justice-and-strong-institutions.

World's Largest Lesson. 2022. https://worldslargestlesson.globaltgoals.org.

World's To Do List. 2022. https://www.project-everyone.org/case-study/the-worlds-to-do-list

XPrize Foundation. 2022. $100M Prize for Carbon Removal. https://www.xprize.org/prizes/elonmusk.

ZISPOA website, https://www.zispoa.info.

Zubaşcu, F. 2021. Don't Mind the Gap: 'Innovation Cohesion' Is New Route to Bridge East-West Divide. Science Business. https://sciencebusiness.net/news/dont-mind-gap-innovation-cohesion-new-route-bridge-east-west-divide.

Appendix A

Committee on Operationalizing Sustainable Development: Biographical Information

E. WILLIAM COLGLAZIER (Co-Chair) is Editor-in-Chief of *Science & Diplomacy* and Senior Scholar in the Center for Science Diplomacy at the American Association for the Advancement of Science (AAAS). He served as the fourth Science and Technology Adviser to the Secretary of State from 2011 to 2014. From 1994 to 2011, he was Executive Officer of the National Academy of Sciences and the National Research Council where he helped to oversee the studies that provide independent, objective scientific advice on domestic and international public policy issues. Dr. Colglazier received his Ph.D. in theoretical physics from Caltech in 1971, and prior to 1994 worked at the Stanford Linear Accelerator Center, the Institute for Advanced Study in Princeton, the Center for Science and International Affairs at Harvard's Kennedy School of Government, and the University of Tennessee. He is a past chair of the Forum on Physics and Society and the Committee on International Scientific Affairs of the American Physical Society (APS) and a fellow of the AAAS and APS. In 2015 he received the Joseph A. Burton Forum Award from the APS for "outstanding contributions to the public understanding or resolution of issues involving the interface of physics and society" and the Order of the Rising Sun, Gold Rays with Neck Ribbon from the Government of Japan for "contributing to science and technology exchange and mutual understanding between Japan and the United States." From 2016 to 2018, Dr. Colglazier served as co-chair of the Ten-Member Group appointed by the UN Secretary General to advise on science, technology, and innovation to achieve the 17 Sustainable Development Goals of the UN 2030 Agenda.

CHERRY MURRAY (NAS/NAE) (Co-Chair) is Professor of Physics and Deputy Director for Research, Biosphere 2 at The University of Arizona. Her current research interests include policy, research, development, education, and innovation to sustain human civilization on future Earth. From 1978 to 2004, Dr. Murray held a number of research positions, which culminated in the Senior Vice Presidency of Physical and Wireless Research, at Bell Laboratories, Lucent Technologies, formerly AT&T Bell Laboratories and previously Bell Telephone Laboratories, Inc. She then served at Lawrence Livermore National Laboratory as Deputy Director for Science and Technology from 2004 to 2007, and as Principal Associate Director for Science and Technology from 2007 to 2009. She was Dean of Harvard University's School of Engineering and Applied Sciences from 2009 to 2014. Dr. Murray served as the Director of the U.S. Department of Energy's Office of Science from 2015 to 2017. She served as President of the American Physical Society in 2009, on the National Commission on BP Deepwater Horizon Oil Spill and Offshore Drilling in 2010, and on numerous National and American Academy, American Association for the Advancement of Science, American Physical Society, Department of Energy, and Department of Commerce committees, and currently serves as chair of the board of the Okinawa Institute of Science and Technology Graduate University and as a director of the American Academy of Arts and Sciences. In 2019, she was elected, representing the U.S. National Academy of Sciences, as co-chair for science of the Inter Academy Partnership, a partnership of more than 140 national and regional science, engineering, and medical academies dedicated to providing independent and credible advice to policy makers and to strengthening the role of academies nationally, regionally, and globally. In 2021 she was appointed co-chair of the Ten-Member Group of the Technology Facilitation Mechanism of the UN. Dr. Murray received her B.S. and Ph.D. in physics from the Massachusetts Institute of Technology.

ERIN BROMAGHIM serves as the Deputy Mayor of International Affairs in the Office of Los Angeles Mayor Eric Garcetti, focused on bringing global opportunities to Angelenos and connecting Los Angeles to economic and cultural partners around the world. She leads a team with deep expertise on international trade and investment, international relations, educational and cultural exchange, the Sustainable Development Goals (SDGs) and the green economy, city diplomacy, gender equity, and major global events including the 2026 FIFA World Cup and the 2028 Olympic and Paralympic Games. Most recently, Ms. Bromaghim served as the Director of Olympic and Paralympic Development as part of the Mayor's International Affairs team, where she led the city's planning for the Olympic and Paralympic Games in 2028 and its enduring benefits for all Angelenos. This legacy includes her work as the Conrad N. Hilton Foundation Fellow, using the framework of the UN Agenda 2030 to align, measure, and track the city's progress toward the 17 SDGs. Ms. Bromaghim also served as a visiting senior fellow on city and state diplomacy with the Truman Center for National Policy, exploring the ways in which collaboration between local and federal actors can advance U.S. foreign policy objectives. Ms. Bromaghim previously spent 14 years as a senior civilian with the U.S. Department of Defense, where she man aged interagency defense, intelligence, special operations, and security reform efforts. She entered federal civil service as a Presidential Management Fellow with the U.S. Navy, later working for the U.S. Air Force, the Office of the Secretary of Defense, and NATO. She holds degrees from Wake Forest University and Georgetown University, as well as a certificate in advanced project management from Stanford University.

HARINI NAGENDRA is Director, Research Centre and Professor and Lead, Centre for Climate Change and Sustainability at Azim Premji University. Over the past 25 years, she has been at the leading edge of research examining conservation in forests and cities of South Asia from the perspective of both landscape ecology and social justice. For her interdisciplinary research and practice, she has received a number of awards including the 2009 Cozzarelli Prize from the U.S. National Academy of Sciences (with Elinor Ostrom), the 2013 Elinor Ostrom Senior Scholar award, and the 2017 Clarivate Web of Science award. Her publications include the books *Nature in the City: Bengaluru in the Past, Present and Future* (Oxford University Press, 2016) and *Cities and Canopies: Trees in Indian Cities* (Penguin, 2019, with Seema Mundoli) and more than 150 peer-reviewed publications, including in *PNAS (Proceedings of the National Academy of Sciences of the United States of America)*, *Nature*, *Nature Sustainability*, and *Science*. Professor Nagendra writes a monthly column "The Green Goblin" in the Deccan Herald newspaper, and is a well-known public speaker and writer on issues of urban sustainability in India. She is also the author of a historical mystery fiction series set in 1920s colonial Bangalore. Professor Nagendra has been a lead author on the Fifth Assessment Report of the UN Intergovernmental Panel on Climate Change, and is a past Science Committee member of DIVERSITAS and the Global Land Programme. She is an Associate Editor of *Global Environmental Change*, serves on the advisory board of the World Resources Institute Ross Centre for Sustainable Cities and the European Institute of Technology's Climate Knowledge and Innovation Centre, and engages with international science and policy through her involvement as a steering committee member of the Future Earth Programme for Ecosystem Change and Society and the Future Earth Urban Knowledge Advisory Network.

NEBOJSA NAKICENOVIC is the Executive Director of The World In 2050 (www.TWI2050.org). He is Deputy Chair of the Group of Chief Scientific Advisors to the European Commission, was the Deputy and Acting Director General of the International Institute for Applied Systems Analyses, and was tenured Professor of Energy Economics at Vienna Technology University. Among other positions, Dr. Nakicenovic has been a member of the Earth League; Earth

Commission of the Global Commons Alliance; Multi-stakeholder Technical Group of Advisors on Sustainable Development Goal 7; Scientific Advisory Boards of the Potsdam Institute from Climate Impact Research; Fondazione Eni Enrico Mattei; Japanese Institute of Environmental Studies; German Aerospace Center; Renewable Energy Policy Network for the 21st Century; OMV Advisory Group on Sustainability; and Climate Change Centre Austria. He has also been a Technology and Innovation Advisor to the government of Montenegro. He serves on many editorial boards of peer-reviewed journals, including *Technological Forecasting and Social Change*; *Climate Policy, Energy Policy, Institution of Civil Engineers*; *Current Opinion in Environmental Sustainability*; *Energy Sector Management*; *Ecosystem Health and Sustainability*; *Scientific World Journal*; *Environmental Innovation and Societal Transitions*; and the *Energy Strategy Reviews*. Dr. Nakicenovic's research interests include the long-term patterns of technological change, economic development and response to climate change, and, in particular, the evolution of energy, mobility, and digital technologies. Dr. Nakicenovic holds bachelor's and master's degrees in economics and computer science from Princeton University and the University of Vienna, where he also completed his Ph.D. He also holds an Honoris Causa Ph.D. degree in engineering from the Russian Academy of Sciences.

ILONA OTTO is a professor in societal impacts of climate change at the Wegener Center for Climate and Global Change at the University of Graz. She leads a research group focusing on social complexity and system transformation. The group's ambition is to use complex science theory and novel research methods to analyze social dynamic processes and interventions that are likely to spark rapid social changes necessary to radically transform the interactions of human societies with nature and ecosystem services in the next 30 years. Dr. Otto is a social scientist by training. She uses various research methods including social surveys, case studies, behavioral experiments, and simulations in analyzing problems related to global environment changes, development, adaptation, and sustainability. Dr. Otto is a principal investigator in an EU Horizon 2020 Project CASCADES: Cascading Climate Risks: Towards Adaptive and Resilient European Societies. She also coordinates a Climate Knowledge and Innovation Community (KIC) Project REBOOST: A Boost for Rural Lignite Regions. She led a chapter on human health in the World Bank report *Turn Down the Heat: Confronting the New Climate Normal*, contributed to the 2020 *UN Emission Gap Report*, and led the *Report on Modelling the Impact of Climate Change on Poverty at a Subnational Scale* that was contracted by the World Bank. Dr. Otto received her Ph.D. in resource economics from the Humboldt University of Berlin.

ALFRED WATKINS is the Founder and Chairman of the Global Solutions Summit (GSS), which focuses on innovative business models and financing strategies to promote the large-scale deployment in emerging markets of commercially viable, financially sustainable development solutions for potable water, renewable energy, information and communications technology, health care, housing, sustainable agriculture, food processing, and manufacturing. Prior to founding the GSS, Dr. Watkins worked for more than 23 years at the World Bank as the World Bank's Science and Technology Program Coordinator and head of the World Bank's Science, Technology and Innovation Global Expert Team. He has extensive on-the-ground experience leading science, technology, and innovation capacity-building programs and projects in Africa, Asia, and various countries in the Former Soviet Union, including Russia, Ukraine, Kazakhstan, and Latvia. Dr. Watkins organized two World Bank Global Forums on Science, Technology and Innovation Capacity Building for Sustainable Development (one in 2007 and another in 2009). Before joining the World Bank, he also served as a legislative assistant/staff economist in the U.S. Congress and was an assistant professor at the University of Texas at Austin. In March 2017, Secretary-General António Guterres appointed him to a 3-year term on the Governing Council of the Technology Bank for the Least Developed Countries (LDCs). In January 2020, the Secretary-General reappointed him to a second 3-year term. The main objective of the Technology Bank, which was formally established by the UN General Assembly in December 2016, is to help LDCs identify and deploy the technology to foster inclusive sustainable growth and achieve the Sustainable Development Goals. Dr. Watkins received his Ph.D. in economics from the Graduate Faculty of the New School University.

Appendix B

Workshop Agendas

INFORMATION GATHERING WORKSHOP 1: LOCAL STRATEGIES
April 18 and 21, 2022
All times are US Eastern Daylight Time

Workshop Objective: Gather information on positive case studies for operationalizing sustainable development at a local level.

Day 1: Monday, April 18, 2022

10:00 am   **Opening Remarks and Introductions**
E. William Colglazier, American Association for the Advancement of Science (Co-Chair)
Cherry Murray (NAS/NAE), University of Arizona (Co-Chair)
Franklin Carrero-Martínez, National Academies of Sciences, Engineering, and Medicine

10:05 am   **Panel I: Sustainable and Equitable Food Systems**
Moderator: Alfred Watkins, Global Solutions Summit
- Eduard Müller Castro, University for International Cooperation
- Alison Grantham, Grow Well Consulting, LLC
- Sabine O'Hara, University of the District of Columbia
- Ravi Kanbur, Food System Economics Commission and Cornell University

11:25 am   BREAK

11:40 am   **Panel II: Education and Capacity Building**
              Moderator: E. William Colglazier, American Association for the
              Advancement of Science
              • Sarah Mendelson, Carnegie Mellon University
              • Carol O'Donnell, Smithsonian Institution
              • Mark McCaffrey, The Long Game
              • Meghna Tare, The University of Texas, Arlington

1:00 pm   Open Session Adjourns

Day 2: Thursday, April 21, 2022

10:00 am   **Welcome and Introductions**
              E. William Colglazier, American Association for the Advancement
              of Science (Co-Chair)
              Cherry Murray (NAS/NAE), University of Arizona (Co-Chair)
              Franklin Carrero-Martínez, National Academies of Sciences,
              Engineering, and Medicine

10:05 am   **Panel III: Urbanization**
              Moderator: Harini Nagendra, Azim Premji University
              • Worajit Setthapun, Chiang Mai Rajabhat University
              • Lykke Leonardsen, City of Copenhagen
              • Marc Weiss, Global Urban Development
              • Susan Parnell, University of Cape Town (by video)

11:25 am   BREAK

11:40 am   **Panel IV: Localization of the Sustainable Development Goals
              and Indigenous Knowledge Networks**
              Moderator: Cherry Murray (NAS/NAE), University of Arizona
              • Emilia Saiz, United Cities and Local Governments
              • Anthony Pipa, Brookings
              • Celeste Connors, Hawai'i Green Growth
              • Ashish Kothari, Kalpavriksh (by video)

1:00 pm   Open Session Adjourns

INFORMATION GATHERING WORKSHOP 2: GLOBAL STRATEGIES
Part 1: May 4, 2022
All times are US Eastern Daylight Time

Virtual Side Event of the 7th Multi-stakeholder
Forum on Science, Technology and Innovation for the
Sustainable Development Goals (STI Forum)

Meeting Objective: Gather information on positive case studies for operationalizing sustainable development at a global level.

10:30 am   **Opening Remarks and Introductions**
Cherry Murray (NAS/NAE), University of Arizona (Committee Co-Chair)
Franklin Carrero-Martínez, National Academies of Sciences, Engineering, and Medicine

10:35 am   **Framing Remarks**
Jan Marco Müller, Science and Technology Advisor, European External Action Service

10:45 am   **Panel: Science, Technology, and Innovation Cooperation**
Moderator: Erin Bromaghim, City of Los Angeles
- Klaus Tilmes, Senior Policy Advisor and Development Consultant
- Niki Frantzeskaki, Utrecht University
- Atsushi Sunami, Sasakawa Peace Foundation
- Jean-Paul Adam, United Nations Economic Commission for Africa

12:00 pm   **Side Event Conclusion**

INFORMATION GATHERING WORKSHOP 2: GLOBAL STRATEGIES
Part 2: May 16–17, 2022
All times are US Eastern Daylight Time

- **Workshop Objective**: Gather information on positive case studies for operationalizing sustainable development at a global level.

Day 1: Monday, May 16, 2022

10:00 am   **Opening Remarks and Introductions**
E. William Colglazier, American Association for the Advancement of Science (Co-Chair)

Cherry Murray (NAS/NAE), University of Arizona (Co-Chair)
Franklin Carrero-Martínez, National Academies of Sciences, Engineering, and Medicine

10:05 am   **Panel I: Decarbonization**
Moderator: Cherry Murray (NAS/NAE), University of Arizona (Co-Chair)
- Erin Burns, Carbon180
- Kathleen Draper, International Biochair Initiative
- Prateek Bumb, Carbon Clean
- Stephanie Arcusa, Arizona State University

11:25 am   BREAK

11:40 am   **Panel II: Science and Peace**
Moderator: Ilona Otto, University of Graz
- Jannie Lilja, Stockholm International Peace Research Institute
- Aparna Basnyat, United Nations Development Programme
- Rachel Locke, University of San Diego
- Roman Słowiński, Polish Academy of Sciences

1:00 pm   Open Session Adjourns

Day 2: Tuesday, May 17, 2022

10:00 am   **Opening Remarks and Introductions**
E. William Colglazier, American Association for the Advancement of Science (Co-Chair)
Cherry Murray (NAS/NAE), University of Arizona (Co-Chair)
Franklin Carrero-Martínez, National Academies of Sciences, Engineering, and Medicine

10:05 am   **Panel III: Financing for Sustainable Development**
Moderator: Erin Bromaghim, City of Los Angeles
- Marianna Koval, New York University
- Adam Roy Gordon, United Nations Global Compact
- Lauren Muusse, World Benchmarking Alliance

11:30 am   Open Session Adjourns

Appendix C

Biographies of Workshop Speakers

INFORMATION GATHERING WORKSHOP 1: LOCAL STRATEGIES

Panel I: Sustainable and Equitable Food Systems

ALISON GRANTHAM is a scientist who brings a methodical, analytical, and quantitative approach to her work with nonprofits, private-sector businesses, and foundations to improve our food system through her practice, Grow Well Consulting. Current and recent projects include improving climate impacts of pasture-raised poultry, greenhouse gas, water and waste footprinting for an indoor agricultural business, a national urban food waste and food insecurity analysis and report, global seafood traceability to support food safety and sustainability outcomes, and a FLAG sector scope 3 engagement for an international environmental nongovernmental organization. Prior to launching Grow Well, she led Food Systems R&D and then Food Procurement at Blue Apron, overseeing food sourcing and procurement and implementing a national program to increase employee access to surplus product, as well as local communities through partnerships with Feeding America. There, she also served on the National Academies' Ad Hoc Reducing Food Loss and Waste Committee. Previously, she led research at the Rodale Institute, including all aspects of organic and sustainable agriculture research. Dr. Grantham holds a dual-title Ph.D. in ecology and biogeochemistry from The Pennsylvania State University and B.A. summa cum laude in biological sciences and environmental studies from Mount Holyoke College.

RAVI KANBUR is T.H. Lee Professor of World Affairs, International Professor of Applied Economics, and Professor of Economics at Cornell University.

Dr. Kanbur researches and teaches in development economics, public economics, and economic theory. He is well known for his role in policy analysis and engagement in international development. He is Co-Chair of the Food Systems Economics Commission. He has served on the senior staff of the World Bank including as Chief Economist for Africa. Dr. Kanbur has also published in the leading economics journals, including *Journal of Political Economy*, *American Economic Review*, *Review of Economic Studies*, *Journal of Economic Theory*, and *The Economic Journal*. The positions he has held include Chair of the Board of United Nations (UN) University-World Institute for Development Economics Research, member of the Organisation for Economic Co-operation and Development (OECD) High Level Expert Group on the Measurement of Economic Performance, President of the Human Development and Capability Association, President of the Society for the Study of Economic Inequality, member of the High Level Advisory Council of the Climate Justice Dialogue, Co-Chair of the Scientific Council of the International Panel on Social Progress, and member of the Core Group of the Commission on Global Poverty.

EDUARD MÜLLER CASTRO is the Founder and President of the University for International Cooperation and Chair-holder for the UN Educational, Scientific and Cultural Organization (UNESCO). The University for International Cooperation is a global organization of educational innovation for an inclusive knowledge society, seeking a regenerative and evolutionary development to a complex, diverse, and changing world. Dr. Müller's career spans more than 35 years and he has published more than 50 papers and book chapters. One of his most notable works was a chapter titled "December 22nd," in a book titled *Global Chorus*. Dr. Müller has spoken for numerous organizations including The Earth Charter, the World Meteorological Organization, and the Latin American Parliament. His greatest achievement thus far has been having more than 5,000 students from 58 different countries graduate from his University. Dr. Müller attributes his success to his experience living in different countries. Dr. Müller is also an accomplished photographer.

SABINE O'HARA is Distinguished Professor and Ph.D. Program Director of the College of Agriculture, Urban Sustainability and Environmental Sciences (CAUSES) at the University of the District of Columbia, which is the only public university in Washington, DC, and the only exclusively urban Land-grant University in the United States. Prior to her current appointment, she served as the founding Dean of CAUSES and led the University's efforts to build a cutting-edge model for urban agriculture and urban sustainability that integrates training in the agricultural, environmental, and health sciences with the practical aspirations of students and residents to embark on successful careers in the green innovation economy. Dr. O'Hara's work has focused consistently on the quality of life and economic opportunity of local communities through multidimensional intellectual, social, and physical capacity development. A foundation of her work is her belief that education should not only answer our questions, but also question our answers. This search for new answers has guided her work as the 10th President of Roanoke College in Virginia; Provost of Green Mountain College in Vermont; faculty member at Rensselaer Polytechnic Institute, in Troy, New York; and Executive Director of the Council for International Exchange of Scholars (CIES), which administers the prestigious Fulbright Scholar Program. Dr. O'Hara holds master's and doctoral degrees in agricultural economics and environmental economics from the University of Göttingen, Germany. She holds an affiliated faculty appointment with the Working Group on Institutional Analysis of Socio-Ecological Systems at Humboldt University in Berlin, Germany; is the past President of the International Society for Ecological Economics (ISEE) and the U.S. Society for Ecological Economics (USSEE); is a member of the International Advisory Board of King Abdulaziz University in Jeddah; and served on the editorial board of several academic journals.

Panel II: Education and Capacity Building

MARK MCCAFFERY is currently involved with a project called The Long Game, which aims to involve schools around the world in community-based climate action projects. Previously, he was founder of the UN Climate Change (UNCC) community ECOS (Education, Communication, and Outreach Stakeholders). Having helped establish a local watershed focused network two decades ago (BASIN, the Boulder Area Sustainability Information Network) and a national digital library and online community (CLEAN, the Climate Literacy and Energy Awareness Network), Mr. McCaffery assisted in initiating the climate and energy literacy frameworks and authored *Climate Smart & Energy Wise*, published in 2014 by Corwin. Relocating from North America to Central Europe to pursue international collaboration and engagement opportunities, he assisted in the launch of the Institute for Sustainable Development in Budapest and has consulted with Climate-KIC, the largest EU public-private partnership focused on climate solutions, and various UN organizations, such as the Food and Agriculture Organization of the United Nations and UNCC: Learn. He has previously served as Programs and Policy Director at the National Center for Science Education (NCSE) and Associate Scientist at CIRES, the Cooperative Institute for Research in Environmental Sciences at the University of Colorado at Boulder, where he developed climate change and energy literacy education programs.

SARAH MENDELSON is Distinguished Service Professor of Public Policy at Carnegie Mellon University (CMU) and Head of CMU's Heinz College in Washington, DC. Ambassador Mendelson previously served as the U.S. Representative to the Economic and Social Council at the UN until January 20, 2017. Confirmed by the Senate in October 2015, she was the USUN (United States Mission to the United Nations) lead on international development, human rights, and humanitarian affairs. There she oversaw campaigns to get country-specific resolutions passed in the General Assembly and to get nongovernmental organizations, including the Committee to Protect Journalists, accredited to the UN. She led efforts to elevate the issue of combating human trafficking and was Senior Lead for the President's Summit on Refugees. Prior to her appointment as Ambassador, she served as a Deputy Assistant Administrator at the U.S. Agency for International Development (USAID) from 2010 to 2014 where she was the Agency lead on democracy, human rights, and governance. A long-time policy entrepreneur, she has spent more than two decades working on development and human rights as a scholar and practitioner including in Moscow with the National Democratic Institute, on the faculty of the Fletcher School at Tufts University, and over a decade as senior adviser and inaugural Director of the Human Rights Initiative at the Center for Strategic and International Studies. There she also worked as a senior fellow in the Russia and Eurasia Program where she oversaw focus groups, public opinion surveys, and social marketing campaigns in Russia on a range of issues. A member of the Council on Foreign Relations and the author of more than 70 scholarly and public policy publications, Ambassador Mendelson received her B.A. in history from Yale University and her Ph.D. in political science from Columbia University.

CAROL O'DONNELL is the Senior Executive and Director of the Smithsonian Science Education Center, an organization of the Smithsonian Institution dedicated to transforming K-12 Education through Science™ in collaboration with communities across the globe. In her role at the Smithsonian (a nonprofit with quasi-governmental status), Dr. O'Donnell serves as the U.S. representative on the Global Council of the InterAcademy Partnership Science Education Programme, an appointment by the U.S. National Academies of Sciences, Engineering, and Medicine; and she serves on the UN Broadband Commission Working Group on School Connectivity: Hybrid Learning. Dr. O'Donnell also represents the Smithsonian on the Subcommittee on Federal Coordination in STEM Education, which advises and assists the Committee on STEM Education of the Office of Science and Technology Policy of the Executive Office of the President. In her role on the Program Committee for the International Dialogue on STEM Education, Dr. O'Donnell co-authored the position paper on "STEM Education for Sustainable Development" (http://bit.ly/3a3ObkS). Prior to joining the Smithsonian, Dr. O'Donnell was a group leader at the U.S. Department of Education, supporting states' and districts' implementation of Elementary and Secondary Education Act programs; she also oversaw the Cognition and Student Learning research grant program of the Institute of Education Sciences. A former K-12 teacher and curriculum developer, Dr. O'Donnell is still in the classroom today, serving on the part-time faculty of the Physics Department at The George Washington University, where she earned her doctorate. Her TedX talk demonstrates her passion for integrating digital and physical interactions in science classrooms.

MEGHNA TARE is the first Chief Sustainability Officer for The University of Texas, Arlington (UTA). Ms. Tare works collaboratively to foster partnerships among academic, research, and operational departments at UTA, and to address opportunities to promote sustainability in several areas including energy efficiency, resource conservation, waste management, transportation, education, outreach, community engagement, supporting and encouraging student initiatives, and implementing an interdisciplinary and sustainability-focused curriculum. Ms. Tare serves and represents UTA on several advisory boards including the National Academy of Sciences Board on Higher Education and Workforce Development—Policy and Global Affairs, Association for the Advancement of Sustainability in Higher Education, and Local Government for Sustainability (ICLEI), and she is a fellow at Hunt Institute for Engineering and Humanity at Southern Methodist University. She has also served on the Advisory Committee for the City of Dallas Environmental and Climate Action Plan, National Academy of Sciences Airport Cooperative Research Program panel, and the Water Resource Council of North Central Texas Council of Governments. She has spearheaded launching a Regional Center of Expertise for Education in Sustainable Development Goals in North Texas, a program of the UN University, and the Institute for Sustainability and Global Impact at UTA. She is a Tedx speaker, was featured as the Women in CSR by Triple Pundit, and was awarded Women of the Decade in Corporate Social Responsibility by the Women Economic Forum and 2020 Women in Sustainability—Transformational Leader by Wells Fargo. Ms. Tare graduated with an M.B.A. in sustainable management, M.S. in environmental science, and M.S. in chemistry.

Panel III: Urbanization

LYKKE LEONARDSEN is the Program Director for Resilient and Sustainable City Solutions for Copenhagen, Denmark. She has more than 25 years of experience in various fields of urban development, including local regeneration projects, international urban policies, and communication. Since 2008 she has worked for the Technical and Environmental Administration in charge of making Copenhagen more blue and green—tasked with water management and green infrastructure planning. It was as part of this work that the city's Climate Change Adaptation plan was developed. She was the key driver for the work on climate change adaptation, the development of the adaptation plan and the cloudburst management plan, and the first steps toward a storm surge plan in Copenhagen and is currently focusing on sharing her city's climate results with other cities around the world.

SUSAN PARNELL is a Global Challenges Research Professor in the School of Geography at the University of Bristol and Emeritus Professor at the African Centre for Cities at the University of Cape Town. She has held previous academic positions at Wits University and the University of London. She was a Leverhulme Visiting Professor at University College London in 2011–2012, Emeka Anyaoku

Visiting Chair University College London in 2014–2015, and Visiting Professor at LSE (London School of Economics and Political Science) Cities in 2017–2018. She has been actively involved in local, national, and global urban policy debates around the 2030 Sustainable Development Goals (SDGs) and is an advocate for better science policy engagement on cities. She is the author of numerous peer-reviewed publications that document how cities, past and present, respond to policy change. Her most recent books include the co-authored *Building a Capable State: Post Apartheid Service Delivery* (Zed, 2017) and the co-edited *The Urban Planet* (Cambridge, 2018). Dr. Parnell is currently on the Board of the International Institute for Environment and Development, serves as a member of the African Centre for Cities Advisory Board, and had previously served on several nongovernmental organization structures.

WORAJIT SETTHAPUN is the Founder and currently the Dean of the Asian Development College for Community Economy and Technology (adiCET), Chiang Mai Rajabhat University. adiCET was established to facilitate sustainable development of the local community. Dr. Setthapun established and managed the Chiang Mai World Green City (CMGC) as the living laboratory for renewable energy and green technologies developed from more than 70 sustainable energy/environment projects. CMGC focuses on the implementation of Community Power for small rural communities. The aim is to develop an affordable/appropriate sustainable energy system to enhance the livelihood and occupation of the community. Dr. Setthapun is the head of the Graduate Program in Community Energy and Environment and oversees the Renewable Energy Research and Training Center in CMGC. Dr. Setthapun has created short course and training programs that trained more than 5,000 participants on community sustainable development, renewable energy, energy and environmental conservation, and the sufficiency economy concept. Dr. Setthapun was the ASEAN (Association of Southeast Asian Nations)-U.S. Science and Technology Fellow and worked at the Ministry of Energy, Thailand's Decentralized Community Power Project. In addition, Dr. Setthapun is the recipient of the 2016 ASEAN-US Women in Science Award for thematic area of sustainable energy.

MARC A. WEISS is Chairman and CEO of Global Urban Development (GUD), an international policy organization and professional network of more than 700 leaders and experts 60 countries. He is a Lead Partner of the UN-Habitat World Urban Campaign, Board Member of IHC Global, and International Visiting Professor at the Federal University of Rio Grande do Sul (UFRGS) in Porto Alegre, Brazil, where he coordinates the Porto Alegre Sustainable Innovation Zone (ZISPOA). He previously served as Public Policy Scholar and Editor of Global Outlook at the Smithsonian Institution's Wilson Center; Coordinator of the 1998 Strategic Economic Development Plan for Washington, DC, and Chairman of the NoMa Metro Station Corporation; Special Assistant to the Secretary of the U.S. Department of Housing and Urban Development (HUD) and HUD Liaison to the President's Council on Sustainable Development in the Clinton Administration; Associate Professor and Director of the Real Estate Development Research Center, Acting Director of the Ph.D. Program in Urban Planning, and Adjunct Professor of International and Public Affairs at Columbia University; and Deputy Director of the California Commission on Industrial Innovation. He is author or co-author of many books, articles, and reports, including *The Rise of the Community Builders* and *Real Estate Development Principles and Process*. He has been a senior adviser on metropolitan economic strategy, sustainable innovation, and inclusive prosperity for cities, counties, regions, and states/provinces throughout the world, including Australia, Brazil, Canada, China, Czech Republic, Germany, India, Italy, Morocco, Panama, Singapore, South Africa, Spain, Sweden, UK, United States, and Virgin Islands. He earned an MCP and Ph.D. in city and regional planning from the University of California, Berkeley and a B.A. with honors in political science from Stanford University. He also attended the London School of Economics.

Panel IV: Localization of the Sustainable Development Goals and Indigenous Knowledge Networks

CELESTE CONNORS is the Executive Director of Hawai'i Green Growth. Ms. Connors has 20 years of experience working at the intersection of economic, environment, energy, and international development policy. Before joining Hawai'i Green Growth, she was CEO and co-Founder of cdots development LLC, which works to build resilient infrastructure systems and services in vulnerable communities. Ms. Connors previously served as the Director for Environment and Climate Change at the National Security Council and National Economic Council in the White House where she helped shape the Administration's climate and energy policies, including the SDGs. Prior to joining the White House, Ms. Connors served as a diplomat in Saudi Arabia, Greece, and Germany. She also held positions at the U.S. Mission to the UN, served as the Climate and Energy Advisor to the Under Secretary for Democracy and Global Affairs at the U.S. Department of State, and worked for the City of New York. Ms. Connors is a senior adjunct fellow at the East-West Center and was a faculty member at the Johns Hopkins University School of Advanced International Studies in the Energy, Resources and Environment Program. She holds an M.Sc. in development studies from the University of London's School of Oriental and African Studies and a B.A. in international relations from Tufts University. Ms. Connors has served on numerous boards including her current service on Hawaiian Electric Industries, the Global Island Partnership, and the Institute for Sustainability and Resilience at the University of Hawaii and Icebreaker One. She previously served on the Board of America's Service Commissions and the IUCN World Conservation Congress National Host Committee, and was a Term Member on the Council on Foreign Relations.

ASHISH KOTHARI is the founder of Kalpavriksh, an Indian nonprofit organization working on environmental and social issues at local, national, and global levels. A graduate in sociology, Mr. Kothari has taught environment at the Indian Institute of Public Administration in the 1990s and has been guest faculty at several universities, institutes, and colleges. He has served on the Board of Directors of Greenpeace International and Greenpeace India. Mr. Kothari has served on the Indian Government's Environmental Appraisal Committee on River Valley Projects and Expert Committees to formulate India's Biological Diversity Act and National Wildlife Action Plan. He coordinated India's National Biodiversity Strategy and Action Plan process and was co-coordinator of the Activist-Academic Co-Generation of Knowledge on Environmental Justice (www.acknowlej.org) global project. Mr. Kothari has been active with a number of people's movements, including Narmada Bachao Andolan (Save Narmada Movement) and Beej Bachao Andolan (Save the Seeds Movement). He helps coordinate the Vikalp Sangam (Alternatives Confluence, www.vikalpsangam. org) process in India and the Global Tapestry of Alternatives (https://globaltap-estryofalternatives.org). He helps run the website and associated list of Radical Ecological Democracy (www.radicalecologicaldemocracy.org). Mr. Kothari is the author or editor of more than 30 books (including *Churning the Earth: Making of Global India*; *Alternative Futures: India Unshackled*; and *Pluriverse: A Post-Development Dictionary*) and more than 400 articles.

ANTHONY PIPA is a senior fellow in the Center for Sustainable Development, housed in the Global Economy and Development program at Brookings. He studies place-based policies to improve social progress in the United States and globally, including through use of the SDGs at the local level. He is also considering the future of U.S. multilateral aid and the applicability of lessons from international development to improving rural development in the United States. Mr. Pipa has more than 25 years of executive experience in the philanthropic and public sectors addressing poverty and advancing inclusive economic development. During the Obama administration, he served as Chief Strategy Officer at the U.S. Agency for International Development and held multiple senior policy positions at the Agency. He served as U.S. Special Coordinator for the Post-2015 Agenda at the Department of State, leading the U.S. delegation at the UN to negotiate and adopt the SDGs. Prior to his government service, he directed the nongovernmental organization Leaders Forum at Harvard University and was the founding CEO of the Warner Foundation, a family foundation in North Carolina focused on improving economic opportunity and race relations. He helped launch Foundation for Louisiana in the aftermath of Hurricane Katrina, and has played a principal role in the start-up of several philanthropic ventures focused on addressing poverty and improving distressed communities. He serves on the Board of Directors of StriveTogether and the Advisory Council of the Center for Disaster Philanthropy. He has published articles, book chapters, and opinion pieces on local implementation of the SDGs, the effectiveness of place-based policies, multilateral aid, philanthropic effectiveness, financial innovations, and policies to strengthen resilience and prosperity. He attended Stanford University, graduated from Duke University, and earned an M.P.A. at the Harvard Kennedy School.

EMILIA SÁIZ is Secretary General of United Cities and Local Governments (UCLG). Ms. Sáiz is a jurist by profession and has devoted her professional life to promoting the role of local governments in development as well as fostering relations between cities and their associations worldwide. She started her journey as local government international advocate in the founding organization of UCLG, IULA (International Union of Local Authorities) in 1997. She has led programs dedicated to institutional capacity building and decentralized cooperation. She has actively promoted women empowerment, social inclusion, and international partnerships. She played a critical role in setting up the Global Taskforce of Local and Regional Governments and has followed and represented local and regional governments in iconic international processes such as the Rio and Beijing + 20 as well the Climate Agreement, the SDGs, and Hábitat III. She has further been deeply involved in the institutional development of UCLG into a network of networks. During her work for the World Organization she was appointed to numerous panels both in her personal and professional capacity. Ms. Sáiz has also been involved in Cities Alliance, the General Alliance of Partners, and the Cities Programme of the Global Compact.

INFORMATION GATHERING WORKSHOP 2: GLOBAL STRATEGIES

Panel I: Science and Technology Cooperation (STI Forum Side Event)

JEAN-PAUL ADAM is the Director for Technology, Climate Change and Natural Resources Management in the UN Economic Commission for Africa (since January 2020). He previously served in several Cabinet positions in the Government of Seychelles including Minister of Finance, Trade and the Blue Economy (2015–2016) where he negotiated a debt for climate change adaptation swap in 2015, which placed 30 percent of Seychelles oceanic space under protection, and launched the process for Seychelles to become the first issuer of a Blue Bond. He also served as Seychelles' Minister of Foreign Affairs (2010–1015) and Minister of Health (2016–2019).

NIKI FRANTZESKAKI is a Chair Professor of Regional and Metropolitan Governance and Planning, Section Spatial Planning, Geosciences Faculty, Utrecht University, the Netherlands. Her expertise is on urban transitions and their governance and planning, with a focus on sustainability, resilience, livability, and more specifically strengthening people-nature interactions in the urban environment. She has a rich international research experience with a portfolio of ongoing projects in Australia, Canada, and the United States. She has been a Highly Cited Researcher awardee from Clarivate Analytics in 2020 and 2021, putting her in the top 1 percent of researchers globally in the cross-field of social sciences and ecology. From 2019 to 2021, she was a research professor and Director of the Centre for Urban Transitions in Swinburne University of Technology, Melbourne, Australia. From 2010 to 2019, she was an associate professor at the Dutch Research Institute for Transitions with Erasmus University Rotterdam. She has published nearly 100 peer-reviewed articles and released four books on urban sustainability transitions in 2017, 2018, and 2020. She has also edited 18 special issues in top-ranked journals on sustainability, sustainability transitions, and urban governance.

JAN MARCO MÜLLER is Science & Technology Advisor in the Strategic Policy Planning Division of the European External Action Service. Previously, he was Head of Directorate Office/Coordinator for Science to Policy and Science Diplomacy, International Institute for Applied Systems Analysis. He was also the Policy Officer for International Relations in the Joint Research Centre, and before that Assistant to the EU Chief Scientific Adviser, Professor Dame Anne Glover. In that role he managed Professor Glover's office, assisted her in her daily tasks, and coordinated her relations with internal and external stakeholders. Professor Glover, who previously served as Chief Scientific Adviser (CSA) to the Scottish Government, reported directly to the Commission President and had the mandate to provide independent expert advice on any aspect of science, technology, and innovation. Before his assignment to the CSA's Office, Dr Müller worked for three years as Assistant to the Director-General of the Joint Research Centre, the European Commission's in-house science service. Previous roles include Head of Business Development & Public Relations of the UK Natural Environment Research Council's Centre for Ecology & Hydrology; Programme and Communications Manager of the European Commission's Institute for Environment and Sustainability in Ispra (Italy); Assistant to the Scientific Director of the Helmholtz Centre for Environmental Research in Leipzig (Germany); and Secretary of the Partnership for European Environmental Research. He also served for four years on the Scientific Advisory Board of the French national environmental research centre CEMAGREF (now IRSTEA). Holding a Ph.D. in geography, Dr. Müller has often worked with Latin American countries, in particular Colombia. He speaks six languages and has a keen interest in developing the science-policy interface.

ATSUSHI SUNAMI is the President of The Sasakawa Peace Foundation, as well as the Ocean Policy Research Institute of The Sasakawa Peace Foundation. He is also Director of the SciREX center and Executive Advisor to the President at the National Graduate Institute for Policy Studies and Guest Professor at the Research Organization for Nano & Life Innovation at Waseda University. He is currently serving as a member of the Basic Policy Group under the Committee on National Space Policy in the Cabinet Office, and as Chair of the Space Utilization Promotion Round-table under the Minister for Space Policy in the Cabinet Office.

In addition, he is a member of the Innovation Strategy for Security and Safety at the Cabinet Office and on the Advisory Board for the Promotion of Science and Technology Diplomacy in the Ministry of Foreign Affairs, Japan. He holds a B.S.F.S. from Georgetown University and an M.I.A. and Ph.D. in political science from Columbia University.

KLAUS TILMES is a Senior Policy Adviser and former World Bank Director with more than 30 years of international experience in development policy, strategy development at the global, country, and sectoral levels, and program implementation. Since leaving the World Bank, Mr. Tilmes has provided policy advice to international organization, governments, and private-sector companies on economic trends, human development, data policies, and technology strategies. As senior expert, he works closely with the African Center for Economic Transformation and the UN organizations on science, technology, and innovation. At the World Bank, Mr. Tilmes worked most recently with the Office of the President to develop the institution's strategy on emerging technologies and scaling adoption through financial assistance, policy advice, and public-private partnerships. He served as Director of the Trade and Competitiveness Global Practice, overseeing operations across Sub-Saharan Africa, the Middle East, and North Africa and leading global expert teams on trade competition policy, innovation, and entrepreneurship. During his tenure at the World Bank, Mr. Tilmes also held positions as Operation and Strategy Director for Finance and Private Sector Development; Knowledge Strategy Advisor; and Manager at the Independent Evaluation Group. He earned an M.P.A. in public administration from Harvard University as a McCloy Scholar and a master in economics from the University of Mannheim.

Panel II: Decarbonization

STEPHANIE ARCUSA is a postdoctoral researcher at the Center for Negative Carbon Emissions at Arizona State University. She received a bachelor's degree in earth sciences from University College Cork, Ireland, a master's degree in climate science from the University of Bern, Switzerland, and a doctorate in climate and environmental change from Northern Arizona University, USA. As a paleoclimatologist, Dr. Arcusa reconstructed changes in the environment (e.g., wildfire, floods, dust) as it responds to climatic changes (e.g., temperature, precipitation) through time. In this work, Dr. Arcusa also saw the fingerprint of human activities on the climate and the environment, which led her to decide to transition her career to help stop the change in climate occurring today due to human activities. Her work now consists of developing ways to halt climate change. Over her academic training, Dr. Arcusa has led or been part of various projects that further emission reduction at the local, city, and higher education levels. She is exploring three related ideas to closing the carbon loop and supporting the development of a new carbon economy. First, she is developing a framework for the certification of carbon sequestration by exploring what certificates are, how they work, and how they can guarantee safe, equitable, and successful sequestration. As part of the work, Dr. Arcusa is developing ways to include carbon-intensive sectors of the economy into the framework so that they can also transition to a circular carbon economy. Finally, Dr. Arcusa is taking part in an effort led by the Los Alamos National Laboratory to plan the decarbonization of the intermountain west region of the United States through a place-based approach focusing on hydrogen, biomass, and carbon capture technologies.

PRATEEK BUMB co-founded Carbon Clean with Aniruddha Sharma in 2009. Mr. Bumb is the principal innovator of carbon capture technologies. He is responsible for developing and delivering Carbon Clean's technology roadmap and leading its project engineering team. He is also a member of the board. His expertise involves innovation and implementation of carbon capture technology from idea to commercial stage. To date, he has international patents on the carbon capture process and solvents and has published papers at several international conferences and in journals on carbon capture and storage technology. Mr. Bumb is skilled in areas such as gas, sustainable development, environmental issues, the Clean Development Mechanism, and environmental awareness. He is a strong entrepreneurship professional and graduated from Indian Institute of Technology, Kharagpur.

ERIN BURNS is the Executive Director of Carbon180, a climate nongovernmental organization focused on the full range of carbon removal solutions. There she leads the organization in working with policy makers, entrepreneurs, and scientists to reach a just and equitable economy that removes more carbon than it emits. Previously, she worked in the Senate where she handled energy, environment, labor, and agricultural issues, including staffing for the Energy and Natural Resources Committee and the Public Lands Subcommittee. She also worked at Third Way, a DC-based think tank, managing carbon capture and removal, innovation, and other clean energy policy development and advocacy. A native of southern West Virginia, throughout her career she has worked on issues related to coal worker and coal community transitions. She holds a degree in cultural anthropology from Carnegie Mellon University.

KATHLEEN DRAPER is a member of the International Biochair Initiative (IBI) Board and Chair of IBI's Information Hub. She is also the U.S. Director of the Ithaka Institute for Carbon Intelligence. The Institute is an open source network focusing on beneficial carbon sequestration strategies, which simultaneously provide economic development opportunities both in the developed and developing world. She is an editor and writer for *The Biochar Journal*, sponsored by the Ithaka Institute. Ms. Draper also works with various universities and individuals on projects that are investigating the use of biochar in cement and other building and packaging products to develop products with lower embodied carbon that can be made from locally available organic waste. She has written extensively about various topics related to biochar and is a co-author of the book *Terra Preta: How the World's Most Fertile Soil Can Help Reverse Climate Change and Reduce World Hunger*.

Panel III: Science and Peace

APARNA BASNYAT is Senior Research and Policy Advisor, SDG16 at the UN Development Programme (UNDP). Ms. Basnyat has more than 15 years of experience working on rule of law, human rights, and governance policy and programming at the country, regional, and global levels. She has supported UNDP country offices in the Asia Pacific and the Arab States regions, most recently as the Policy and Programme Specialist with the Rule of Law, Security and Human Rights Team in the Crisis Bureau where she focused on engagement on access to justice and SDG 16. She has advised on country programming on rule of law across different development settings, supported inter-agency cooperation and partnerships with civil society on access to justice, and led research and policy development on justice and human rights. Ms. Basnyat holds a masters in development studies from the London School of Economics and Political Science and an undergraduate degree in international relations from Tufts University.

JANNIE LILJA is Director of Studies, Peace and Development at Stockholm International Peace Research Institute. Until recently she served as a senior fragility specialist with the World Bank's Fragility Conflict Violence Group. Prior to joining the World Bank she was a Swedish diplomat and researcher with the Department of Peace and Conflict Research at Uppsala University. Her areas of expertise cover peacebuilding, war and peace dynamics, negotiations, and development in conflict settings. She has published several articles and book chapters on these topics. Dr. Lilja has contributed to the development of peacebuilding policy, including SDG 16, in her former capacity as lead on security and development in the Swedish Ministry for Foreign Affairs. She has also served as Swedish Delegate to the UN Human Rights Council in Geneva. Dr. Lilja received a B.Sc. and M.Sc. in international economics and business from the Stockholm School of Economics, a M.Sc. in development studies from the London School of Economics, and a Ph.D. in peace and conflict research from Uppsala University.

RACHEL LOCKE is Director of the Violence, Inequality and Power Lab at the Kroc Institute for Peace and Justice (IPJ), University of San Diego. Ms. Locke has extensive experience delivering evidence-based violence prevention solutions to some of the most challenging contexts while simultaneously advancing policy for peace. Prior to joining IPJ, Ms. Locke was Head of Research for violence prevention with the Pathfinders for Peaceful, Just and Inclusive Societies at New York University's Center on International Cooperation. In this capacity,

Ms. Locke led coalition building and evidence curation with the UN, bilateral governments, the African Union, civil society, and others to explore the challenge of delivering the 2030 Agenda targets for peaceful societies (SDG 16.1). Earlier in her career, Ms. Locke served as Senior Policy Advisor with the U.S. Agency for International Development (USAID) where she developed and represented agency-wide policy on issues concerning conflict, violence, and fragility. She also led USAID research and policy on crime, conflict, and fragility and worked extensively on program design, implementation, and evaluation primarily in Africa. After leaving USAID, Ms. Locke launched a new area of work for the National Network for Safe Communities at John Jay College of Criminal Justice, bridging effective violence reduction approaches from the United States to municipalities globally. Among other initiatives, Ms. Locke launched a three-year effort across two states and five municipalities in Mexico at a time of exceptionally high violence. Ms. Locke holds a master's in international affairs from Columbia University Graduate School of International and Public Affairs. She has also published a variety of articles and other works focusing on violence prevention, humanitarian aid, conflict, and transnational organized crime.

ROMAN SŁOWIŃSKI is a Professor and Founding Chair of the Laboratory of Intelligent Decision Support Systems at Poznan University of Technology, Poland and Vice President of the Polish Academy of Sciences, elected for the term 2019–2022. Dr. Słowinński is a member of Academia Europaea and a fellow of the Institute of Electrical and Electronics Engineers, the International Rough Set Society, the Institute for Operations Research and the Management Sciences, and the International Federation for Information Processing. In his research, he combines Operational Research and Artificial Intelligence for Decision Aiding. He is a recipient of the EURO Gold Medal by the European Association of Operational Research Societies (1991), and Doctor HC of Polytechnic Faculty of Mons (Belgium, 2000), University Paris Dauphine (France, 2001), and Technical University of Crete (Greece, 2008). In 2005 he received the Annual Prize of the Foundation for Polish Science—the highest scientific honor awarded in Poland. Since 1999, he has been the principal editor of the *European Journal of Operational Research* (Elsevier), a premier journal in operational research.

Panel IV: Financing for Sustainable Development

ADAM ROY GORDON is Engagement Director for Global Compact Network USA and in this role leads the work of the UN Global Compact in the United States, including operations, participant management, and programming. He is a contributor to *The Atlantic* and was named to the Environment + Energy Leader 100 in 2019. Previously, Mr. Gordon worked at CDP, supporting the integration of climate change, water, and deforestation disclosure into corporate performance. He was an EDF Climate Corps Fellow at Colgate-Palmolive Company and has diverse experience in sustainability that ranges from advising the government of Montenegro on green building policy to founding New York City's first commercial composting waste hauler. He holds a B.A. from Oberlin College and an M.S. in sustainability management from Columbia University.

MARIANNA KOVAL is the Director of the Invest NYC SDG Initiative at the New York University's (NYU's) Stern Center for Sustainable Business, where she manages a team that is engaging the private sector to drive financing toward creating a more sustainable, inclusive, and resilient New York City (NYC). Together, using the UN SDGs as a guiding framework, the team is developing investable projects to advance NYC's sustainability goals in waste, food and health, climate resilience, renewable energy, the built environment, and sustainable mobility. A lawyer, with more than 30 years of experience working in environmental sustainability, public policy, and government in NYC, Ms. Koval has successfully taken on complex multi-stakeholder projects and difficult policy challenges, and brought conflicting interests together to create concrete projects. She served as the senior advisor to the Commissioner of NYC Environmental Protection, building green infrastructure policy and partnerships in a $2.4 billion stormwater management program. She created the first NRDC-DEP partnership to develop a private market for green infrastructure, a project that she later managed from the NYU Stern Center for Sustainable Business, which produced an influential report, *Catalyzing Green Infrastructure on Private Property: Recommendations for a Green, Equitable, and Sustainable New York City.* Ms. Koval was President of the Brooklyn Bridge Park Conservancy for more than a decade, where she helped develop the vision, attract the initial $280 million in funding, and construction of this major city park. She holds an M.P.A. from the Kennedy School of Government at Harvard University, a J.D. from Fordham University Law School, and an A.B. from Princeton University.

LAUREN MUUSSE co-leads the World Benchmarking Alliance's (WBA's) investor engagement strategy. She connects investors and investor-focused platforms to the work of the WBA. This expands to developing new modes of collaboration with other financial institutions and ESG (environmental, social, and governance) service providers. She sees these actors as key catalysts for transformation across the SDGs. She is particularly interested in systems thinking and connecting the dots across transformation areas to the work of the financial industry. Ms. Muuse also contributes to thinking and collaboration via different work streams within the WBA on the topic of human rights. She previously lead the human rights work at ING Bank that focused on the bank's own work and also broader engagement and collaboration in order to affect systems change on human rights due diligence. She holds an M.A. in political science and indigenous studies and a B.A. in native studies from the University of Alberta. She strives to bring what she has humbly learned about decolonisation to all her work.